Dairy Farm Business Management

Dairy Farm Business Management

by

P. Venkateshwara Rao

BIOTECH

2025

Biotech Books

ISBN10 81-7622-195-3
ISBN13 978-81-7622-195-5

Published by : **BIOTECH BOOKS**
4762-63/23, Ansari Road, Darya Ganj,
New Delhi - 110002
Phones : 011 4300 3222
Email : biotechbooks@yahoo.co.in

PRINTED IN INDIA

Preface

The dairy farm business is one of the largest sub-industries of the whole agricultural process, generating revenues worth $500 billion worldwide. Involving the production of milk, butter, cheese and other milk products using livestock and milk processing technologies, dairy farm business management entails a whole deal of practices which are pertinent to the whole process of dairying as an ever-growing industry, especially considering the fact that the demand for milk products has reached an all-time high. In such a scenario, it becomes incumbent upon the dairy farmer to not just look after the management of the dairy farm itself, but also check upon its viability as a dependent source of income and business opportunity.

The present book goes beyond the conventional discussion of dairying as a farm operation, and delves into the business aspect of the field. Delineating the commercial aspects of dairying, it makes assessment of the principles, trends and developments which are moulding the dairying business, and the prospects offered by it. In addition, space has been given to analysing the growth of the business, and making evaluation of the challenges and advantages which surround it. It depicts at length about the dairy business management with latest guidelines and is a must for every body whosoever is interested directly or indirectly in dairy business management.

P. Venkateshwara Rao

Contents

1

Introduction to Dairy Farming

Dairy farming has been part of agriculture for thousands of years, but historically, it was usually done on a small scale on mixed farms. Specialist scale dairy farming is only viable where either a large amount of milk is required for production of more durable dairy products such as cheese, or there is a substantial market of people with cash to buy milk, but no cows of their own. Centralized dairy farming as we understand it primarily developed around villages and cities, where residents were unable to have cows of their own due to lack of grazing land. Near the town, farmers could make some extra money on the side by having additional animals and selling the milk in town.

The dairy farmers would fill barrels with milk in the morning and bring it to market on a wagon. Before mechanization most cows were still milked by hand. At milking time they brought the vacuum pump, and the automatic milking machine. The first milking machines were an extension of the traditional milk pail. The early milker device fit on top of a regular milk pail and sat on the floor under the cow. Following each cow being milked, the bucket would be dumped into a holding tank. This developed into the Surge hanging milker. Prior to milking a cow, a large wide leather strap called a surcingle was put around the cow, across the cow's lower back.

The milker device and collection tank hung underneath the cow from the strap. This innovation allowed the cow to move around naturally during the milking process rather than having to stand perfectly still over a bucket on the floor. Surge later developed a vacuum milk-return system known as the Step-Saver, to save the farmer the trouble of carrying the heavy steel buckets of milk all the way back to the storage tank in the milkhouse. The system used a very long vacuum hose coiled around a receiver cart, and connected to a vacuum-breaker device in the milkhouse.

Following milking each cow, the hanging milk bucket would be dumped into the receiver cart, which filtered debris from the milk and allowed it to be slowly sucked through the long hose to the milkhouse. As the farmer milked the cows in series, the cart would be rolled further down the centre aisle, the long milk hose unwrapped from the cart, and hung on hooks along the ceiling of the aisle.

The next innovation in automatic milking was the milk pipeline. This uses a permanent milk-return pipe and a second vacuum pipe that encircles the barn or milking parlour above the rows of cows, with quick-seal entry ports above each cow. By eliminating the need for the milk container, the milking device shrank in size and weight to the point where it could hang under the cow, held up only by the sucking force of the milker nipples on the cow's udder. The milk is pulled up into the milk-return pipe by the vacuum system, and then flows by gravity to the milkhouse vacuum-breaker that puts the milk in the storage tank.

The pipeline system greatly reduced the physical labour of milking since the farmer no longer needed to carry around huge heavy buckets of milk from each cow. The final innovation in automatic milking was the milking parlour, which streamlined the milking process to permit cows to be milked as if on an assembly line, and to reduce physical stresses on the farmer by putting the cows on a platform slightly above the person milking the cows to eliminate having to constantly bend over.

Milking parlors allowed a large concentration of technical equipment to gather in one place, which permitted automatic milk take-off devices. Before this, milking was not entirely automatic, and each cow needed to be monitored so that the milker could be removed when the cows were almost done lactating. Leaving the milker on too long following lactation could lead to health problems such as mastitis.

Early Milk Preservation Methods

Keeping milk cool helps preserve it. When windmills and well pumps were invented, one of its first uses on the farm besides providing water for animals was for cooling milk, to extend the storage life before being transported to the town market. The naturally cold underground water would be continuously pumped into a tub or other containers of milk set in the tub to cool after milking. This method of milk cooling was extremely popular before the arrival of electricity and refrigeration.

When refrigeration first arrived, the equipment was fairly small and did not have the ability to rapidly cool the large volume of milk that was entering the storage tank in a short period of time. This problem was resolved through the development of the ice bank. This is a double-walled tank design where water and cooling coils fill the space underneath and around the milk tank above. All day long, the small compressor and cooling system slowly draws heat out of the water, while a second pump continuously circulates the water around the coils. Ice eventually builds up around the coils, until it reaches a thickness of about three inches surrounding each pipe, and the cooling system shuts off.

When the milking operation starts only the milk agitator and the water circulation pump blowing water across the ice and the steel walls of the tank are needed to rapidly reduce the incoming milk to a temperature below 40 degrees. But because the ice is not permitted to build up until it touches the milk storage tank, the milk does not get cold enough to also freeze. This cooling method worked well for smaller dairies up to about 40 cows, but for large numbers of animals a better system was needed to rapidly cool the incoming warm milk. This is usually done using a device known as a plate chiller, which is a heat exchanger.

Alternating stainless steel plates cause the milk to flow in a thin sheet across the plates, while cold water is circulated in a thin sheet on the other side of the plates. Flattening out the milk flow permits quick, even cooling for all the milk, compared to a round tube where the centre core does not cool as rapidly as the walls. The plate chiller has high cooling demands, and for many farms this involves a step back into the past, back to the days of windmills and milk-can cooling, except now a large volume of naturally cold underground water is continuously streamed through the plate chiller to quickly bring down the milk down to the temperature of the underground water at about 50 degrees F. The water is usually not just dumped back into the ground again, but reused for washing and other purposes.

But the milk still is not as cold as it needs to be, so the milk storage tank is still used to do further cooling, to bring the milk down to 40 degrees. But with the development of high-power 3-phase electrical service, ice-bank chillers are typically no longer used. Instead the milk storage tank is a direct-cooling system with cooling coils embedded in the walls of the tank, that quickly pull the heat out and dump it across a large array of possibly several different high-horsepower compressors and condensing units. Once the milk has achieved 40 degrees F after milking is finished, only one or two cooling units need to run occasionally to maintain the correct temperature.

Milking Operation

Hand Milking Processes

Until the late 1800s, the milking of the cow was done by hand. In the United States, several large dairy operations existed in some northeastern states and in the west, that involved as many as several hundred cows, but an individual milker could not be expected to milk more than a dozen cows a day. Smaller operations predominated. Milking took place indoors in a barn with the cattle tied by the neck with ropes or held in place by stanchions. Feeding could occur simultaneously with milking in the barn, although most dairy cattle were pastured during the day between milkings. Such examples of this method of dairy farming are difficult to locate, but some are preserved as a historic site for a glimpse into the days gone by. One such instance that is open for public tours is at Point Reyes National Seashore.

With the availability of electric power and suction milking machines, the production levels that were possible in stanchion barns increased but the scale of the operations continued to be limited by the labour intensive nature of the milking process. Attaching and removing milking machines involved repeated heavy lifting of the machinery and its contents several times per cow and the pouring of the milk into milk cans. As a result, it was rare to find single-farmer operations of more than 50 head of cattle.

Modern Milking Parlour Operations

Farmers use any number of styles of milking parlors to milk dairy cattle. Many older farms use flat-barn parlors, where the milker and cow are at the same level and the milker bends down to apply the milking machine to the cow. More modern farms use recessed parlors, where the milker stands in a recess such that his arms are at the level of the cow's udder.

Recessed parlors can be herringbone, where the cows stand in two angled rows either side of the recess and the milker accesses the udder from the side, parallel, where the cows stand side-by-side and the milker accesses the udder from the rear or, more recently, rotary, where the cows are on a raised circular platform, facing the centre of the circle, and the platform rotates while the milker stands in one place and accesses the udder from the rear. There are many other styles of milking parlors which are less common.

In herringbone and parallel parlors, the milker generally milks one row at a time. The milker will move a row of cows from the holding yard into the milking parlour, and milk each cow in that row. Once all or most of the milking machines have been removed from the milked row, the milker releases the cows to their feed. A new group of cows is then loaded into the now vacant side and the process repeats until all cows are milked. Depending on the size of the milking parlour, which normally is the bottleneck, these rows of cows can range from four to sixty at a time.

In rotary parlors, The cows are loaded one at a time onto the platform as it slowly rotates. The milker stands near the entry to the parlour and puts the cups on the cows as they move past. By the time the platform has completed almost a full rotation, another milker or a machine removes the cups and the cow steps backwards off the platform and then walks to her feed. Milking machines are held in place automatically by a vacuum system that draws the ambient air pressure down to 15 to 21 pounds of vacuum.

The vacuum is also used to lift milk vertically through small diameter hoses, into the receiving can. A milk lift pump draws the milk from the receiving can through large diameter stainless steel piping, through the plate cooler, then into a refrigerated bulk tank. Milk is extracted from the cow's udder by flexible rubber sheaths known as liners or inflations that are surrounded by a rigid air chamber. A pulsating flow of ambient air and vacuum is applied to the inflation's air chamber during the milking process.

When ambient air is allowed to enter the chamber, the vacuum inside the inflation causes the inflation to collapse around the cow's teat, squeezing the milk out of teat in a similar fashion as a baby calf's mouth massaging the teat. When the vacuum is reapplied in the chamber the flexible rubber inflation relaxes and opens up, preparing for the next squeezing cycle. It takes the average cow three to five minutes to give her milk. Some cows are faster or slower. Slow-milking cows may take up to fifteen minutes to exhaust themselves of milk.

Milking speed is only minorly related to the quantity of milk the cow produces—milking speed is a separate factor from milk quantity; milk quantity is not determinative of milking speed. Because most milkers milk cattle in groups, the milker can only process a group of cows at the speed of the slowest-milking cow. For this reason, many farmers will cull slow-milking cows. The extracted milk passes through a strainer and plate heat exchangers before entering the tank, where it can be stored safely for a few days at approximately 3°C or around 42°F. At pre-arranged times, a milk truck arrives and pumps the milk from the tank for transport to a dairy factory where it will be pasteurized and processed into many products.

Waste Output From Huge Dairies

As measured in phosphorus, the waste output of 5,000 cows roughly equals a municipality of 70,000 people. In the U.S., dairy operations with more than 1,000 cows meet the EPA definition of a CAFO (Concentrated Animal Feeding Operation), and are subject to EPA regulations. For example, in the San Joaquin Valley of California a number of dairies have been established on a very large scale. Each dairy consists of several modern milking parlour set-ups operated as a single enterprise. Each milking parlour is surrounded by a set of 3 or 4 loafing barns housing 1,500 or 2,000 cattle. Some of the larger dairies have planned 10 or more series of loafing barns and milking parlors in this arrangement, so that the total operation may include as many as 15,000 or 20,000 cows.

The milking process for these dairies is similar to a smaller dairy with a single milking parlour but repeated several times. The size and concentration of cattle creates major environmental issues associated with manure handling and disposal, which requires substantial areas of cropland for manure spreading and dispersion, or several-acre methane digesters. Air pollution from methane gas associated with manure management also is a major concern. As a result, proposals to develop dairies of this size can be controversial and provoke substantial opposition from environmentalists including the Sierra Club and local activists. The potential impact of large dairies was demonstrated when a massive manure spill occurred on a 5,000-cow dairy in Upstate New York, contaminating a 20-mile stretch of the Black River, and killing 375,000 fish. On Aug. 10, 2005, a manure storage lagoon collapsed releasing several million gallons of manure into the

Black River. Subsequently the New York Department of Environmental Conservation mandated a settlement package of $2.2 million against the dairy.

Dairy Herd Management

Modern dairy farmers use milking machines and sophisticated plumbing systems to harvest and store the milk from the cows, which are usually milked twice or thrice daily. During the warm months, in the northern hemisphere, cows may be allowed to graze in their pastures, both day and night, and are brought into the barn only to be milked. Many barns also incorporate tunnel ventilation into the architecture of the barn structure. This ventilation system is highly efficient and involves opening both ends of the structure allowing cool air to blow through the building.

Farmers with this type of structure keep cows inside during the summer months to prevent sunburn and damage to udders. During the winter months, especially in northern climates, the cows may spend the majority of their time inside the barn, which is warmed by their collective body heat. Even in winter, the heat produced by the cattle requires the barns to be ventilated for cooling purposes.

Many modern facilities, and particularly those in tropical areas, keep all animals inside at all times to facilitate herd management. Housing the cow can be either loose housed or stalls (called cow cubicles in UK). In the southern hemisphere milking animals are more likely to spend most of their lives outside on pasture. There is little research available on dimensions required for cow stalls, and much housing can be out of date, however increasingly companies are making farmers aware of the benefits, in terms of animal welfare, health and milk production.

The production of milk requires that the cow be in lactation, which is a result of the cow having given birth to a calf. The cycle of insemination, pregnancy, parturition, and lactation, followed by a "dry" period before insemination can recur, requires a period of 12 to 16 months for each cow. Dairy operations therefore included both the production of milk and the production of calves, or veal. As the size of herds has increased, the conditions in which large numbers of veal calves are raised, fed and marketed on larger dairies also have provoked controversy among animal rights activists.

Dairy Farming in the World

In the United States, the top four dairy states are, in order by total milk production, California, Wisconsin, New York, and Pennsylvania. Dairy farming is also an important industry in Florida, Minnesota, Ohio and Vermont. In Pennsylvania, the dairy industry is the number one industry in the state. Pennsylvania is home to 8,500 farms and 555,000 dairy cows. Milk produced in Pennsylvania yields about US$1.5 million in farm income every year, and is sold to various states up and

Table 1: World production not including countries in the European Union.

Rank	Country	Production (MT/yr)
1	India	96.1
2	United States	67.2
3	Russia	32.8
4	Brazil	23.3
5	China	16.8
6	New Zealand	14.6
7	Australia	10.6
8	Mexico	9.8
9	Turkey	9.5
10	Japan	8.4
11	Canada	8.0
12	Argentina	8.0
13	Switzerland	3.9
14	South Africa	2.6
15	South Korea	2.4
16	Norway	1.6

down the east coast. The world's largest exporter of dairy products is New Zealand. Japan is the world's largest importer of dairy products.

The EU is the largest milk producer in the world, with 143.7 million tonnes in 2003. This data, encompassing the present 25 member countries, can be further broken down into the production of the original 15 member countries, with 122 million tonnes, and the new 10 mainly former Eastern European countries with 21.7 million tonnes.

Dairy Competition

Most milk-consuming countries have a local dairy farming industry, and most producing countries maintain significant subsidies and trade barriers to protect domestic producers from foreign competition. In large countries, dairy farming tends to be geographically clustered in regions with abundant natural water supplies (both for feed crops and for cattle) and relatively inexpensive land (even under the most generous subsidy regimes, dairy farms have poor return on capital). New Zealand, the fourth largest dairy producing country, does not apply any subsidies to dairy production.

The milking of cows was traditionally a labour-intensive operation and still is in less developed countries. Small farms need several people to milk and care for only a few dozen cows, though for many farms these employees have traditionally been the children of the farm family, giving rise to the term "family farm". Advances in technology have mostly led to the radical redefinition of "family farms" in industrialised countries such as the United States.

With farms of hundreds of cows producing large volumes of milk, the larger and more efficient dairy farms are more able to weather severe changes in milk price and operate profitably, while "traditional" very small farms generally do not have the equity or cashflow to do so. The common public perception of large corporate farms supplanting smaller ones is generally a misconception, as many small family farms expand to take advantage of economies of scale, and incorporate the business to limit the legal liabilities of the owners and simplify such things as tax management.

Milk production in India is characterised by small rural producers scattered all over the country accounting for about 70% of production. In India, a far larger proportion of milk continuous to be handled by the unorganised sector i.e. 85% comprising enumerable small processors and manufacturers of indigenous milk products. But the main contention in the unorganised sector is the quality, which creates a serious threat to human health. Similar is the case with poultry, where considerable segment is still unorganised. In the absence of organised marketing, the poultry development programme under the cooperative sector is not able to make a dent.

The Government of India has been considering a proposal to provide financial assistance for setting up/modernisation of dairy and poultry sector. The interventions through the scheme involve the generation of self employment and providing infrastructure to the unorganised sector for making improvement in the quality resulting in food safety. It will help in bringing a significant portion of unorganised sector in the ambit of organised sector and will result in increasing the commercial viability of the activities.

The scheme considered and approved by the competent authority, will be implemented through National Bank for Agriculture and Rural Development (NABARD), which will be the nodal agency for the scheme. Now, the undersigned is directed to convey the administrative approval for implementation of the said Central Sector Plan scheme "DAIRY/POULTRY VENTURE CAPITAL FUND" during the 10th Plan period. The guidelines for the approval of the project under the scheme shall be as below:

i) The financial assistance of 50% of the project cost will be provided by Government of India as interest free loan while 40% of the project cost shall be provided by the financing bank at the rate of interest as applicable for agricultural activities and 10% share of the project cost shall be borne by the beneficiary. Besides this, Government of India will also subsidise the interest component applicable for agricultural activities to the extent of 50% in case of regular/timely repayment by the beneficiary.

ii) The Government of India will release its share to NABARD, which will be maintained by them as revolving fund.

iii) The scheme will be extended to agricultural farmers/ individual entrepreneurs and groups of all sections of unorganised as well as organised sector.

iv) The components, which can be funded under the scheme, are given below. However, they may be funded individually or in combination.

References

Aliki, *Milk from cow to carton*, New York: HarperCollins, 1992.

Cobleigh, R., *Handy farm devices and how to make them*, New York: The Lyons Press, 1996.

Davidson, J., & Chase, L. *Farm machinery: Practical hints for handy-men*, New York: The Lyons Press, 1999.

Gunby, L., *Early farm life*, New York: Crabtree Publishing Company, 1992.

Rath, S., *About cows*, Minocqua, WI: Heartland Press, 1987.

2

Housing of Cattles

Open yard housing means leaving the animals free to move from the exercise area to the feeding through to the litter. This solution has been adopted all over the world for all new constructions for young cattle. There are a number of possible choices which fall under this general principle. Basically, they concern the kind of protection afforded the animals, how the various components are arranged and how manure is removed and stored.

Principles of Housing

Housing young cattle in open yards is still a new technique in most of the developing countries. So before making a detailed study of the building techniques involved in this kind of housing, we really must recall to the reader the principles behind this kind of housing and its advantages over conventional, stanchion housing systems. After a quick look at the various systems recommended, we shall point out the basic recommendations to be observed regarding space standards and the various rounds of work to be respected.

The recommendations in this report apply particularly to housing for young cattle, such as feeder bulls and young heifers. But there is no reason why the stockman cannot adapt them, as needed, to other kinds of stock-rearing. Thanks to the experience gained in the Tunisian project, the authors were able to correct and refine a number of construction details. They were also able to identify the most frequently encountered mistakes and problems in this kind of work. This is why certain items have been dealt with in such a detailed way.

There is no question that proper housing is an important factor in successful stock-rearing, though of course it could not alone solve all the problems involved in intensifying animal production. If well-planned housing is to be productive, then there must of course also be balanced feeding of healthy cattle genetically adapted to the specific production specialisation pursued. Open yard housing, despite its many advantages, is not adaptable to certain, exceptional conditions under which young cattle are reared. But when it is possible, and most of the time it is, it can certainly help at the same time to improve working conditions and boost the stockman's income.

Functions of Housing

Before making a detailed study of the open yard housing system (or loose housing), it would be a good idea to review the necessary functions of housing designed to provide shelter for cattle. One reason why loose housing has made such slow progress in the developing countries is that stockmen in these countries are often some what misinformed as to the function which housing for livestock should perform. Stables are often considered prestige items as buildings, and not enough attention is paid to the real needs of livestock-raising.

Stockmen are also often obsessed with the idea of making something durable, a building built to last 20 or 30 years. What they fail to realise is that the production specialisation for which the building was intended might have to be abandoned, and that, in any case, technological advance is such that housing concepts which seem suitable today may tomorrow be completely out of date. Another frequent misconception about stabling is to give undue emphasis to protection of the animals from inclement weather, whilst overlooking the working conditions of the people involved in caring for the stock.

It so happens that domestic animals, especially cattle, have an astonishing capacity to adapt to differing environments.

Often, in fact, excessive concern with protection, such as stables which completely isolate the animal from the outside, can provoke even greater problems than those the building was designed to avoid. It is because a building must satisfy a number of different functions at the same time that the overall concept of a building should not be altered by the desire of the individual stockman to "improve upon" the initial plan without taking the overall concept into account.

Figure 1.

Protection

First of all, housing has to protect the animal against the hostilities of the environment. Climate varying considerably from one region to another, this function can be filled in very different ways.

— *Protection from heat* is of particular importance at certain times of the year in many of the developing countries. But this kind of protection does not justify a heavy outlay for installations—a simple shelter, it does not have to be waterproof but must offer shade during the sunniest part of the day—is, when well-designed, sufficient to ensure the comfort of the animals and proper air circulation. Even a row of trees or hedge, specially planted for this purpose, is sometimes enough.

— *Protection from cold,* when the drop in temperature is not made worse by windchill, is less crucial. A bit more straw in the lying area and a slight increase in the feed ration is enough to help the animals through the winter season. Cattle can withstand temperatures as low as - 10° C without any particular problem.

— *Protection from wind* is essential during the cold season. It may be imperative to build walls, windbreaks or board up an area against the prevailing winds so as to avoid respiratory diseases which are prevalent where herd density is high.

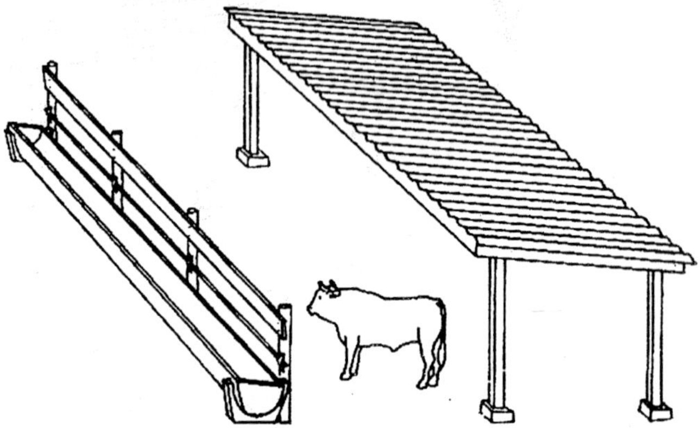

Figure 2.

— *Protection from rain,* which seems so important to stockmen, is only technically and economically justified by rather constant rainfall...However, the production specialisation may have to be temporarily interrupted during the rainy season, as in the case of fattening. In other types of specialisations, it is usually enough to build a roof over the lying area.

— *Protection from mud* is clearly more important. The animals can indeed dry off rather quickly after a shower, but if water has accumulated in the open yard it may remain for several weeks. The proper procedure in this case is to ensure good drainage of the housing area and to adhere strictly to the standard animal density/unit of area.

Feeding

The output from rearing young cattle, whether males for slaughter or females for breeding, can be analysed either from the technical standpoint in terms of mean daily gain or from the economical standpoint in terms of the stockmen's gross margin at the end of the operation. In either case, output depends mainly on feeding.

The success of the operation is dependent upon two factors: the intrinsic quality of the diet, and the volume of feed consumed daily by the animals, which depends on the palatability of the fodder and how it is distributed to the animals.

Accordingly, special attention has to be paid to the arrangement of the feeding-trough in the different housing systems recommended:

- The manger must be such as to allow the animal to eat enough of the foods making up his diet. Feeding-trough capacity must therefore be in line with the kind of diet and how often it is distributed.
- The manger must be designed so that the stockman can easily fill it from outside the housing area without disturbing the animals and as safely and most efficiently for himself as possible.
- The manger must be designed so that all the animals in the lot have access to it at the same time, so that each can eat his share without being crowded out by the more aggressive. It is essential to respect a standard of 50 cm/per head at the trough.
- The manger should be so designed as to cut down on waste and rejected food: withers and protection bars will keep the animals from climbing into the food,

soiling it and scattering the uneaten portion on the ground.

— The manger must be suitable for continuous use throughout the raising or rearing cycle. So it must either be of average dimensions adapted to animals of different weights and/or be provided with a height-regulating device.

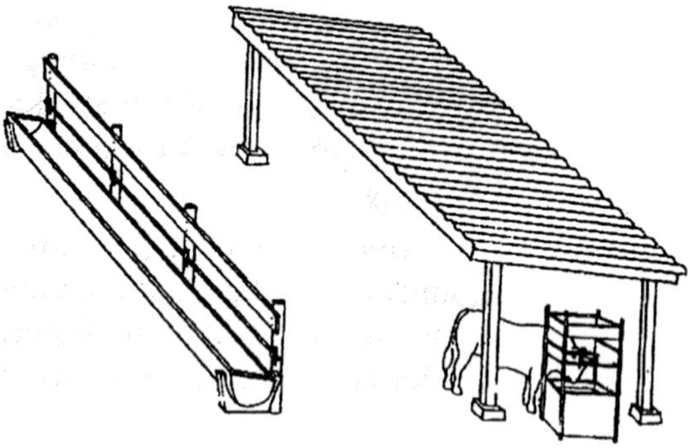

Figure 3.

Watering

Young feeders and breeders have to be able to drink water within their housing area. Water needs vary considerably with the kind of diet (silage fodder is 3/4 ths water in weight compared with 1/10th for concentrated feed), climatic conditions, type of housing and so forth (the environmental surroundings).

And of course water needs vary with the weight of the animals (the average is ten litres of water per 100 kg of live weight per day). In addition to and independently of these quantitative factors, the animals must be able to drink in accordance with their bodily needs, i.e., small amounts several times a day. Naturally, the drinking water has to meet certain chemical and biological standards: excessive dry extract can provoke digestive disorders.

Strong organic odours can cause the animals' refusal to drink. Wherever these conditions for watering the stock are not respected, they will inevitably consume less fodder, and their growth will accordingly be compromised. Unfortunately, in newly set-up installations for feeders and breeders in the developing countries, an all too common occurrence is the flagrant contradiction between intensive-type feed systems, and extensive-type watering systems. So whatever housing system is adapted, it must have a drinking-trough within the housing area where the animals can drink as much and as often as they want. Considering the geographical location of the developing countries, water heating and anti-freeze devices are not usually necessary.

Isolation

One function of housing is, of course, to ensure the isolation of the stock. They ought to live in the area reserved for them. They should particularly be protected from cohabitation with the other farm animals: poultry, dogs, small ruminants, and from cohabitation with man. Thus isolated, the stock remain calm and the risk of accident and contamination is lessened.

Isolation is ensured by a system of fixed and movable fences, doors and separations setting apart lots of animals of similar weights who will remain together during the complete fattening cycle. Indeed, when different lots are mixed together, the fattening process is disturbed by the ensuing agitation, aggressive behaviour, fighting, climbing of one animal on another, and the like. The result is that less food is consumed, weight is lost, and there may be accidents. Fencing can provide isolation, but obviously, in the interests of economy or convenience, all sorts of barriers can be used—brick, beeze-block and even puddled-clay (adobe) walls, bars, branches, pickets, metal tubes and the like.

The only rules which have to be scrupulously observed are those concerning: the height of the obstacle. If it is under 160 cm, even young animals will try to get over it. Possible danger from the obstacle: fencing which could hurt the animals such as certain kinds of farm scraps cobbled together. Strength—sometimes frightened animals throw their whole weight onto the barriers. Visibility—the more the animal is aware of the obstacle the more he respects it. This is why barbed wire is discouraged, particularly where animal density is high, as none of the last three conditions are respected with the use of barbed wire.

When young cattle are isolated, their tranquility is assured, and therefore so is the output of fattening and breeding. In other words, entry into the housing area must always be avoided save for specific operations carried out at specific times and without frightening the animals. The young cattle will be taken out of the housing area as seldom as possible. Such sorties always result in a drop in performance of several days. For instance, efforts should be made to combine weighing operations with medical and sanitary treatments.

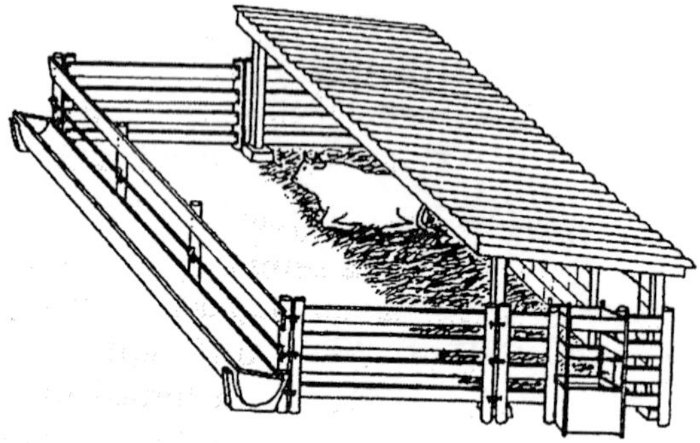

Figure 4.

Comfort of the Animals

The comfort of the animals consists in providing them with the surroundings most propitious to their development.

First of all, the building must be properly ventilated, that is the air must circulate freely enough to avoid accumulations of carbon dioxide and ammonia from the exhalations and excreta of the stock. The building must also allow the animals to get enough rest, when they want it, and moderate exercise as well.

The building must also be easily adaptable to the (possibly mechanized) removal of excreta. Where straw is available and manure is prized, the deep litter system where manure accumulates in the housing area and is removed three or four times a year is to be recommended. But where straw is scarce, or where its use as feed has priority, daily removal of dung is to be recommended. In this case the flooring ought to be concrete (the sloping platform system).

Open yard housing in Europe is frequently based on automatic manure removal by means of overall or partial slatted flooring which allows both dung and urine to pass through the openings by the joint action of trampling and gravity.

The slatted floor technique is not examined in this report for the following reasons:

— Slatted flooring is only applicable to open yards with full roofing, the most costly type.

— Installation costs for slatted flooring and dung channels are substantial.

— The manufacturing of slatted flooring in itself assumes a rather high level of technology.

— Such an installation necessarily demands handling equipment (special pumps), storage equipment (dung channel), transport and spreading (special tank trucks). The introduction of the foregoing would seem to be beyond the technical and financial capacities of stockmen raising young cattle in the developing countries at this time.

Advantages of Open Housing

The advantages of open housing for young cattle, feeders or breeders, are decisive enough for the technique to have spread throughout the cattle-raising countries in the space of the last twenty years.

First of all, we may note that loose housing is the most satisfactory means of filling all the aforementioned housing functions. Before making a detailed examination of the different advantages of loose over stanchion systems, we ought to briefly review the most oft-mentioned objections to the method as expressed by herdsmen:

— *The animals are not sufficiently protected from inclement weather*: Cattle have a noteworthy capacity to adapt to different environments. Loose housing has worked for milk cows and their calves in regions where the temperature drops below 0°C for several consecutive months, and it has worked in regions where temperature are as high, or higher than, 40°C.

— *The animals will fight and hurt themselves*: When minimum space requirements are met, strictly matched lots and dehorning are usually enough to keep the animals quiet.

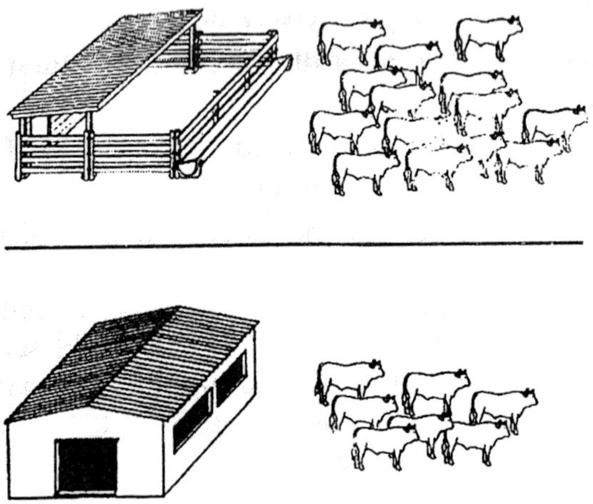

Figure 5.

— *Exercise will make the animals lose weight*: Quite the contrary. We may well assume that moderate exercise maintains the basic bodily functions, particularly appetite. Stockmen usually understand this and let their animals in stanchion systems have a turn in the farmyard each day.

— *Light is harmful to the fattening process*: There is no technical basis for this oft-made remark. Darkness, on the other hand, offers a number of disadvantages.
— *With the deep litter system, the animals cannot be kept clean*: This is not accurate if there is enough straw in the litter—roughly two kg per animal each day. Besides, an animal's growth is in no way compromised by a patch of dung on the flank.

System Requires Less Roofed Area Per Animal Housed

One advantage of the technique is that the young cattle can be housed in installations requiring, in some cases, no roofing at all and in others a covered surface of one/half to one/third that allotted per animal in stanchion stable systems. Indeed, with no-roof open housing there is, of course, no covered area at all. With the combination roof/open yard system, 1.5-2 m of covered area per animal is adequate, and where the area is fully roofed, the standard figure is 3.5-4m per animal.

Stanchion stabling, in contrast, involves a covered area of 4-6 m per animal because the feeding passage (which is outside in open yard housing) and the cleaning passage (there is none in loose housing) have to be covered with no advantage to the stock. This advantage is not only valid for buildings built specially for loose housing for cattle.

The technique can also, in fact, make possible the use of such unused or inadequate existing buildings as sheds, conventional buildings, and the like... For instance, when former stanchion stables are utilised as part of the combined loose housing system, their initial accomodation capacity can be doubled or tripled for the same area of roofing. Roofing is the most expensive item per square metre in construction costs. The financial repercussions of this on cattle housing are easily imaginable. It ought to be pointed out that this trend to reduce the construction portion of housing can also be followed with other types of farm buildings, especially silage fodder silos.

Straw for Litter can be Economised

One great advantage of loose housing, particularly in areas where straw is scarce, or is used for cattle feed, is the fact that this technique requires only 2 kg/animal/day vs at least 4 kg/animal/day for the stanchion stable system under equivalent conditions of hygiene. This is a particularly decisive benefit from the standpoint of straw treatment and utilisation techniques, making possible the upgrading of the nutritive value of this commodity, which can then be used to substitute fodder.

With the deep litter system, where the litter accumulates over a period of months, not only can one save on straw; an excellent quality of manure is obtained which can be spread directly onto the fields after each periodic cleaning of the housing area. This is of particular importance in those areas where stockmen are not yet familiar with manure storage, ripening and utilisation techniques and where, as a result, large quantities of organic fertilizer are lost to agriculture each year.

When cereal straw is not available, it can be replaced by certain agro-industrial wastes: these products, however, must be structurally capable of absorbing moisture from the excreta. Sawdust, leaves and such are examples. When no litter is available in situ at the feeder/breeder locale, the best kind of loose housing is the sloping platform which means little or no litter. The technique consists of building a concrete platform with a maximum width of 4 m and an 8 percent slope towards the manager.

Labour Efficiency

The rapid development of loose housing in most of the cattle-raising countries is basically due to the fact that it improves labour efficiency, which means less manpower, which means a more profitable operation. Conventional stanchion stables often lack a feeding passage, and fodder has to be distributed with a pitchfork whilst moving about among the young cattle. This is disturbing to the animals. Where silage is used, the work is arduous and can even be dangerous for the worker. Moreover, the mangers are not usually capacious enough and so must be filled several times a day.

In contrast, loose housing is so set up that fodder is emptied directly from the wheelbarrow or trailer into the manger from outside the yard. The work is less tiring, easier and quicker. Nor, where running water is available, is watering a chore. The work is limited to transporting the water in a tank truck where there is no running water. Lastly, periodic cleaning out of the litter can be done with the fork-lift and limited to two or three operations per year.

In the final analysis, one worker can only care for 15-25 young bulls in stanchion stable systems. With unmechanised loose housing, the figure rises to 50 and to over 100 where an automatic loading trailer for distributing fodder is available. It should be observed that the improved working conditions possible with open yard housing for young cattle is not merely a quantitative factor. Since the work is less arduous in this kind of installation, more intelligent participation in controlling the animals and in the operation in general can be demanded of the worker.

System is More Economical to Build

Despite all logic, it is a common occurence to see stockmen raising or fattening young cattle and tempted to adopt the most costly, inflexible and unwieldy kinds of solutions to animal housing.

Amortization of livestock buildings represents a fixed cost which can vary substantially, according to:

— initial installation cost,

— utilization method (extent to which building is fully occupied annually),

— useful life and repair and maintenance costs.

Loose housing is advantageous as the initial building costs are much lower than those for conventional stanchion stables. Since the building materials used are light, repair and upkeep of these installations is never very expensive. Moreover, the flexibility of utilisation of certain components of the livestock housing will make it possible to eventually use them for other purpose if breeder/feeder production happens to be abandoned.

The initial building costs of loose housing are lower than stanchion stable building costs, for the following reasons:

- — The extent of covered area: non-existent in no-roof open yards, not extensive in the combined and fully-roofed open yard systems.
- — Except the shelter wall behind the roofed area, no extensive masonry work is called for, as most partitions will simply be fences.
- — The floor covering is reduced to the strict minimum and only in certain parts of the housing area.
- — The materials used in this kind of building can be selected from the most economical available in the region.

Though livestock building costs vary greatly from one region to another, the following estimates can be made: stanchion stable systems cost twice as much as fully-roofed open yards, three times as much as combined systems and five to six times as much as no-roof open yards.

Hygiene Conditions

Open yard housing ensures the animals hygiene conditions much superior to those of stanchion stable systems, especially when the buildings used for the latter system are not well-designed for the purpose. Air circulation is much better than it is in stanchion stable systems. In summer or in hot climates the lack of fresh air and confinement in stables often mean that the air is heavy with ammonia and the temperature so high as to be prejudicial to the animals' health and well-being.

As mentioned before, open yard housing ensures greater comfort for the stock. They can move about freely within the enclosure from the feeding area near the manger to the exercise area and the rest area, where they can lie down and chew their cuds without being disturbed by the rest of the animals in the lot. Contrary to the widely-held idea that leaving animals free to move about means wasting energy and compromising weight gains, it should be emphasized that a number of comparative studies made in different countries show no significant difference in performance between open yard stock and stock housed in properly designed stanchion stable systems.

Classic stables do not usually have facilities for watering the animals. But in loose housing, a permanent drinking-source can be set up in accordance with the animals' needs. The common feed-trough found in loose housing, which can be checked at any time to see if it needs filling, lets the stock feed at their own speed and consume the volume most nearly approaching their own bodily optimum.

Systems of Open Yard Housing

Housing livestock is a technique—it must not be imagined that any solutions suggested would be either universal or permanent.

They cannot be universal: Animal housing has to be adapted to the specific conditions of a region (climate), farm (area, funds, possible use of existing buildings, site conditions) and such technical constraints as the kind of stock-raising system involved, sensitivity of the stock, and what materials are available.

They cannot be permanent: Due to the highly dynamic nature of the technique, any solution proposed at a given moment for a specific animal housing problem may swiftly prove inadequate. The kind of production may have to be modified. There may be some major change in production techniques.

There are many kinds of open-yard systems for young cattle. Detailed studies have been made on some of these, especially in the temperate-zone countries. Studies based in the dry or rainy hot developing countries are, however, less common. There are three kinds of housing described here.

Unroofed Open Yard

Many stockmen find it hard to imagine that stock can be permanently housed in the open and tend to resist the system. It is, however, the economic solution which requires the least investment and can be recommended to stockraisers who have no existing building suited for adaptation for housing but do have a suitable site for open-yard housing.

The critical element in this system is the nature of the soil. Drainage must be perfect so as to avoid the formation of mud which would cut down on cattle performance. Sloping lands with a gradient of 3-10° are preferable. Sandy soils are also usable. Another desirable feature is trees offering protection from the prevailing winds and shade during the hot season.

The system has proven itself time and again in countries like the United States. There, where climate is often less favourable than in many of the developing countries, the bulk of feeders are fattened in open-yard feedlots. So the no-roof open-yard system is to be recommended wherever the conditions are appropriate. Some stockmen think to improve this system by building a shed roof 2-3 m wide over the manger, thinking to thus protect the feed from intemperate weather.

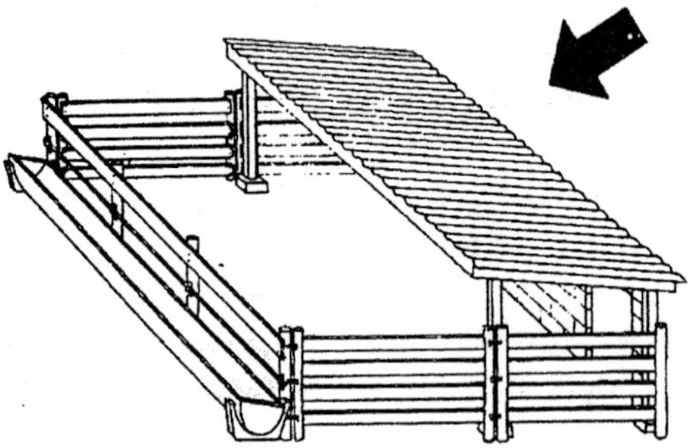

Figure 6.

In addition to the fact that this extra construction adds substantially to the cost of the installation, thus detracting largely from its advantageousness, the benefit is imaginary. Some of the stock will tend to group together under the shelter to get away from the sun and the rain and in so doing will prevent the others from eating. There is also little point in trying to protect concentrated feed, which is consumed in a few minutes. Nor does it seem necessary to protect the coarse fodders against sun and rain.

Combined System

The combined system involves a covered area, the lying area, and an outdoor area for feeding and exercise. This is the system generally favoured by stockraisers, as it has all the advantages of fully covered housing at much lower cost (3-4 m of roofing vs 7-8 m with the fully roofed system). The system must always be oriented in consideration of the direction of prevailing winds. The protected area where the animals can shelter must be against the wind.

Moreover, the combined system offers the great advantage of being easily set up using existing buildings at very little cost. The old building (shed, stanchion stable, lean-to, etc...), is used for the lying area. The feeding trough is built outside and marks the boundary of the unroofed feeding and exercise area. It is true that this kind of arrangement has to take into account certain a prioris concerning the space allotted to each animal, circulation passages and floor covering.

Once this has been taken care of, the technical person in charge of setting up the new housing for young cattle should always find out whether some old, existing building could not be adapted for combined system open housing before recommending the construction of a completely new building of whatever type.

Lastly, stockmen are often tempted to modify the initial design of this kind of housing by building the manger under the covered section and not in the open part of the yard, thinking to thus protect the feed from bad weather as well as protecting the animals. This mistake, which consists of superimposing the feeding area on the resting area, leads to the most harmful consequences.

The animals lying under the shelter prevent the others from getting to the trough to feed. Not only that, the animals eliminate waste mostly when they feed, and so the roofed area of the yard soon becomes an impassable mire. And as has been mentioned, there is little point in protecting feed against sun and rain. So the covered part of the housing is to be set aside as a resting area for the stock.

Full Roofing

As indicated by its name, full roofing is housing completely protected by a roof extending as much as 2 m above the manger. As in the combined system, the building is oriented so that the part opposite the manger is against the prevailing winds. Since the entire structure is covered, the area allotted each animal is substantially reduced with respect to the other two systems.

Covered area per animal, on the contrary, is twice that of the combined system. Roof depth is double that of the preceding system, and so this kind of structure needs a frame with a carrying strength of 6-8 m, implying the use of sturdier, and hence more costly, materials. In addition, fully-roofed housing imposes constraints as to the kind of drinking-trough to be installed. On the other hand, this kind of structure is adaptable to any kind of ground, and concrete flooring is not usually necessary. The sides of the structure are made of movable fences, there are no stationary fences.

However, in view of its costs, this kind of housing will only be recommended under highly specific conditions:

— when the kind of soil or available room make other types of housing impossible;
— when the climate is an unusually rainy one, with prolonged periods of cold and violent winds;
— when constant use of the housing can be planned (near-instant replacement of new lots of stock at the end of the breeding/ fattening cycle).

— when the production specialisation is such as to be able to envision long-term (roughly ten years) utilisation of the housing.

As this latter item is virtually impossible to anticipate, it is always a good idea to employ the most flexible solutions in building fully-roofed systems. The building should be able to be either remodelled or even purely and simply moved somewhere else should the production specialisation be abandoned.

Space Norm Requirements

Respect of space norm requirements is a must; the effectiveness of the housing depends on it. These minimum standards concern the room which must be available to each animal when the entire lot is lined up in front of the manger (trough space) and the space available to each, which is calculated by dividing the total housing area by the number of animals present (ground area).

If these standards are not respected and the building overfilled with stock, a series of negative phenomena which can compromise the technical outcome of the operation will result. These are: nervousness, competitiveness and aggressiveness among the animals, a mud-pit for a floor due to the overpopulation, frequent accidents, destruction of the installations and a drop in performance.

There are no maximum standards other than those dictated by economic necessity, which determines that the amortisation costs of a building used at half capacity doubles per head or per kg of growth compared to a building used at full capacity. The ideal building shape is the square. In fact, a long, rectangular building can create what is called the "corridor effect".

The animals are disturbed in their movements from feeding to lying area and the width of the exercise area is smaller relative to the square shape. This is a situation encouraging aggressiveness and favouring the dominance of the strongest. Housing area can be calculated from room at the trough, ground area and the square shape of the yard, and this gives the number of head which can be accomodated.

It should be pointed out, however, that the standards governing the number of animals in a lot are less rigid than those which have been formulated for length of trough and ground area per animal. Nonetheless, rather large lots can raise problems of supervision and handling. More qualified personnel may be required.

Figure 7.

Room at the Feeding-trough

Minimum space standards per head have been set for each of the three loose housing systems based on observations of studies which have been carried out over a period of several years in many of the cattle-raising countries. They should be thought of as imperatives. If these standards are not respected, the consequences for proper utilisation of the building and sometimes for the technical performances and economic yield of fattening operations can be very serious.

Minimum manger space per head is 50 cm no matter which system is selected. With less space, the risk is arousing competetiveness among the different animals in the lot. The strongest young bulls may prevent the lighter ones from getting to eat. For animals with a liveweight of over 500 kg., trough space per animal should be lengthened to 60 cm.

In most cases, housing capacity will be determined by the length of the trough. It is therefore important to:

— For new housing, avoid any sort of arrangement blocking access to the manger such as setting up drinking-bowls (of any type) less than 3 m away from the manger.

— For readapted old buildings, the manger can be made longer than the building to which it has been annexed. Frontal access to the housing by gates will be avoided whenever possible as these would necessarily shrink the length of the manger.

— In setting up the feeding-trough protection devices, care must be taken to avoid the use of improper or improperly used materials (overly large or badly finished withers bar, manger too low, withers bar too close to manger). These devices must not hinder access to the trough thus making the animals more competetive about getting their feed.

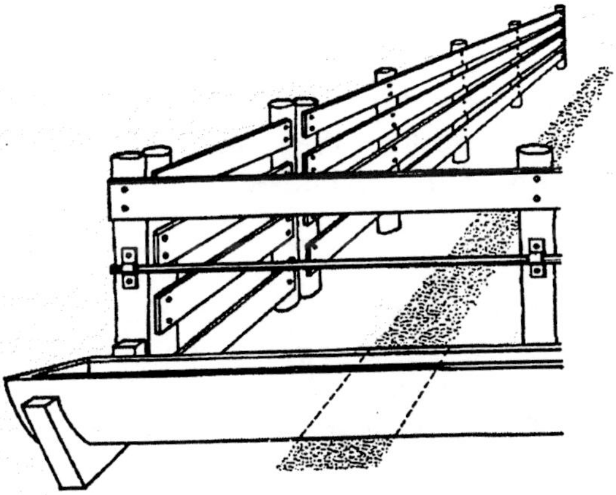

Figure 8.

Ground Area and Lot Size

No-roof open yard

It must be emphasized that the fully open yard system is only to be recommended for very well-drained soils with swift runoff after even the heaviest rains. In this kind of system, the ground area per head must be 10 m, which means that at a rate of 50 cm space at the trough, a well drained fully open yard would have a depth of about 20 m.

Where drainage is not very effective and the water tends to stand in the yard, economy reasons may yet determine the choice of the no-roof open yard system. Should this be the case, it may be necessary to up the figure of ground area per head to 20 or even 30 m. At a rate of 50 cm space at the trough, a badly drained fully open yard may have a depth of 40-60 m. In the first instance, trough length (equal to depth) will be 20 meters and the housing will be able to accomodate 40 head. In the second instance, with the dimensions of the housing area as much as 40 or even 60 m, it will be able to accomodate 80-120 animals.

But it must be stressed that yard capacity cannot be expanded indefinitely. While it is true that each American feedlot can accomodate 150-300 head, the situation is different in the developing countries. Herds of this size require highly specialised supervision, which might well not be available. To facilitate operations and individual observation of the stock, it seems preferable to limit the capacity of completely open yards to 40-50 head maximum.

Combined system

Construction of new housing. The combined system clearly requires less space per head of young cattle. For a roofed area of 1.5-2 m, the outside area can be a mere 2 m, or a total of 3.5-4 m/head. The most desirable shape for the housing being the square, a trough length of 7-8 m equal to the depth of the yard gives an optimum figure of 14-16 head in a lot.

Using these standards, the depth of the roofed part of the housing can also be determined: at a trough space rate of 50 cm/head, the roofing is only 3-4 m deep. Remodelling existing buildings, where the combined system is the best one. Here the problem is different:

When the total depth of the housing is less than 7 meters, the limiting factor becomes the ground area. Even if manger length then exceeds 50 cm/head, the determining factor in calculating the number of head/lot will be ground area. When the depth of the housing is more than 8 meters, manger length becomes the limiting factor determining the number of head/lot, even if the ground area allotted to each is then more than 2 m roofed and 2 m open yard.

When the only communication between the covered and the open area is by gates, which is common in the remodelling of former stanchion stables to open yard systems, housing capacity can be calculated by measuring the covered area and dividing by 2 (2 m roofed area/head). The necessary manger length will be calculated at 50 cm/head.

This manger, which has been set up outside 4 meters from the wall of the old building, will sometimes have to be longer than the building. This is no problem, except that a few more meters of permanent fencing will have to be set up. When this happens the design sometimes has to be changed to a slightly rectangular shape so as to better utilise the covered area. The width is then represented by the long side of the rectangle and the depth by the shorter side. In this case, a combination-type housing can accommodate 20-25 head/lot, assuming, of course, that standards on manger space and ground area allotments per head have been respected.

Full roofing

Full roofing is the system which allows the highest concentrations of young bulls, since only 2.5-3 m of ground area/head is sufficient. But precisely because it is fully roofed, it is also the costliest system. Since the square is the desirable shape for this kind of housing as well, a manger length of 5-6 m equal to the depth of the yard gives an optimum figure of 10-12 animals/lot.

We have to remember that the two-metre shed roof over the front part of the yard (over the feeding-trough) ups the covered area/ head to 3.5-4 m. This shed roof is intended to keep rain and mud out of the covered area rather than to protect the feed in the trough. Under these conditions, the depth of the roof will have to be from 7-8 m.

Where old buildings have been remodelled, it is often easier to adopt this system, especially if the building is closed on all four sides, because it is not usually wide enough to allow enough room for a feeding passage and to house the animals. Likewise, it is often difficult to work out the circulation of the animals. In such a case, a total minimum width of 9 m is imperative.

If the old building is closed on only three sides, such as a tool shed, it can be relatively easy to set up a fully roofed loose housing system if there is sufficient depth (± 5-6 m minimum). But there may be some problem with manure removal if entrance from the sides of the building is not possible. When the sides of the building are walled, then lightweight materials are called for when building the mangers so that they can be removed to let in a tractor and forklift to remove manure. Removal by hand is difficult, though not inconceivable. Minimum height at the lowest part of the shed must be 2.5 m.

Concepts of Livestock Housing Design

Livestock housing design concepts can be very flexible as, for instance, in the selection of materials to be used.

The three basic systems described before also make possible the adaptation of housing to specific, individual farm conditions. However, as we have just seen, there are standards which must be maintained, particularly those concerning space/head, an essential prerequisite. Equally essential are the rounds which must be planned for if utilisation of the building is to be satisfactory. There are three distinct rounds corresponding to activities which must be carried out periodically:

— Feed distribution, which is done twice a day, and drinking-water distribution where a piped-in supply is not available.

— Manure removal, which is done daily where straw is not available and several times a year where the deep litter system is used.

— Moving and handling cattle, especially for weighing, but also for any other operation for which the animals have to be moved out of the yard.

The importance of these rounds or circuits cannot be overstressed. It can even be said that the design concept of the building must be planned around them. None of them should ever be sacrificed with the excuse of expanding the capacity of the housing, for instance, or making use of an already existing fence or shelter. In fact most of the problems that arise with respect to these circuits have to do with remodelling existing buildings for loose housing. In such a case, the solutions found must respect these circuits. Where this cannot be done, the building should not be used for loose housing.

Feeding Passage

When young cattle are fed with fodder supplemented by concentrates, this means four feed distributions a day, not counting the transport of drinking-water for storage-tank watering-troughs. In systems excluding daily manure removal, the bulk of the work to be done in raising young cattle is involved with feed distribution. This is why a feeding passage must be set up and floored, as it is a major factor bearing on labour costs for feeding or breeding.

The feeding passage is laid out parallel to the manger: in such a manner as to allow the circulation of all kinds of mechanical devices. It must be four m. wide and straight, or at least not too curved to let the trailer run along the outer edge of the trough without bumping into anything. Since the feeding passage is the most oft-used circuit, particular attention must be paid to it, especially when remodelling old buildings.

For instance, it is unacceptable to build a manger closer than one meter to a wall, with the idea that the workers can transport the feed by hand. In such a case, the design concept of the remodelled building will just simply have to be modified. The feeding passage must be adaptable for use by trailers in all weather and all seasons. A prerequisite, then, is that the passage be practicable under all circumstances. It is recommended that the ground be blocked up with rubble and then rolled flat with some mechanical device, and paved or asphalted.

Though the supply of cattle feed is not, properly speaking, a housing problem, the problem does arise of storing the feed and the distance which must be crossed between the storage and distribution areas. As a general rule, the feed storage area will be as close as possible to the housing.

Service Passage

When enough straw is available, loose housing is best adapted to the deep litter system, which consists of allowing manure to accumulate, cleaning out the housing only 3 or 4 times a year. What happens is that the dung, which is constantly being trampled by the cattle, is very compressed. After a few months there is a substantial volume of a very heavy, compact product.

Whenever possible, it is highly advantageous to mechanise the operation. Prerequisites for mechanisation are access for a tractor and trailer and a front-lift tractor forklift, and ease of circulation within the housing. For the equipment to circulate within the housing area, there must be a service passage alongside the trough measuring 4 m in width and with no obstacles. The service passage must be continuous, i.e., it must go from one side of the housing to the other so that the tractor can go in one side and come out the other.

There is no special problem with the service passage in the fully open yard system, except for the establishment of movable fences 4 m wide, preferably right behind the manger, where most of the dung collects. In the combined system, the service passage corresponds to the area outdoors and between the manger and the roofed-over area. In the fully-roofed system, the sérvice passage of course runs under the covered area.

Where straw is not available, on the other hand, or has to be used for feed, it is best to use the sloping platform system. This consists of raking the dung daily underneath the manger and out of the yard. The savings on straw are certainly offset by the work involved. It is wearisome for the worker and disturbing to the animals. Even where the sloping platform system is used, it is best to take the precaution of setting up movable fences allowing access to the service passage.

Function of the Access Door

The function of the access door is to let workers and cattle in and out of the housing area.

Young cattle will use the door rather infrequently:

- as they enter the feeding or breeding cycle;
- at weighing;
- when the whole lot is treated, as for vaccinations, parasite protection, etc.
- individually when a sick or wounded animal is moved to the infirmary.
- at the conclusion of the feeding or breeding cycle.

Workers may use this door daily:

- to distribute straw for litter in the lying area for deep litter systems;
- to check on whether the automatic drinking-bowl is working properly and to clean it.

The access door is built into the back of the building, opposite the manger. It thus allows access through the closed-up area of combined or fully-roofed systems. The door is 1 m wide by 2 m high. It is built into one corner of the back wall so as to facilitate the exit of the lot. It will lead directly into the circulation and holding passage which must be annexed to all types of loose housing systems.

Where existing buildings are being remodelled, and especially where former stanchion stables are being used for the roofed area of combined systems, it is often quite impossible to work in an access door. What must be done in such a case is to plan for frontal or lateral access for workers and cattle:

- lateral access by means of movable fences belonging to the service passage, an imperative which must, in all cases, be respected;
- frontal access, which certain kinds of arrangements may make inevitable. In this extreme case, there is the following alternative choice:
  - the installation of movable mangers which will be removed to let the animals by (the workers will have to climb over the fence);
  - a break in cemented mangers (concrete, parpen or irrigation pipe), so as to allow the installation of gates in the front part of the housing area. This solution should be avoided whenever at all possible, because it cuts down on trough length, thus reducing the capacity of the housing.

References

Achilles, W., *Model houses for suckler cow herds: Unconventional ideas are in demand*, Rinderwelt, 1997.

Berg, K., *Houses for self recruiting beef cattle, Norway*, 1996.

Cielejewski, H., *Experience with cold housing for dairy cows, Landtechnik*, 1997.

Fisher, A.D, Crowe, M.A, Prendiville, D.J, Enright, W.J., "Indoor space allowance: effects on growth, behaviour, adrenal and immune responses of finishing beef heifers", *Animal Science: An International Journal of Fundamental and Applied Research*, 1997.

Mader, T.L, Hahn, G.L, Gaughan, J.B., Dahlquist, J.M. "Shade and wind barrier effects on summertime feedlot cattle performance", *Journal of Animal Science*, 1999.

Wassmuth, R, Wallbaum, F, Langholz, H. J., "Outdoor wintering of suckler cows in low mountain ranges", *Livestock Production Science*, 1999.

3

Loose Housing Construction

Housing is made up of a certain number of components, such as the manger, the drinking-trough, the fencing, and so forth, which fill highly specific functions. Each component can be thought of as an identical standard module, unvaried no matter which kind of housing is finally selected for its installation. Strict standards have been formulated for the construction of each of these modules. The builder does, however, have great freedom with regard to building materials, which can be chosen in accordance with availability and price.

The "roof module", for instance, can be tile, eternit, sheet iron, puddled clay or thatch; the frame, wood, iron or reinforced concrete. With "modular" design, standard building components are possible for whichever kind of housing is selected. The manger module, for example, is identical for completely open, combination and fully roofed loose housing.

It is essential to respect these construction standards; sometimes a modification of a few centimetres is enough to prevent one of these components from functioning properly. This is true, for example, of the height of the withers bar and of the width of the holding passage. Any eventual change in these standards should be preceded by a detailed study and lengthy trials. After the range of standard modules has been described would be the right place to examine how each is set up, since a rather strict sequence has to be followed.

Flooring

The selection of the floor covering of the housing area proper and the surrounding space (feeding passage, holding corridor) is very important. On the one hand, the floor covering is an important item in the cost price of construction. On the other, proper utilization of livestock housing depends somewhat on what kind of flooring is selected. In terms of cost, the range of difference between a floor which is merely tamped down and one where concrete is poured can vary from 1 to 20.

In terms of utilization, improper flooring can impede the movements of both man and beast, aggravate the hardship of the actual labour involved, and jeopardize the animals' cleanliness, comfort, and even health.

The major possible alternatives are:

— Ground left as is, perhaps merely well-tamped

— Ground partly or wholly cobbled and flattened with some mechanical device;

— Ground partly or wholly floored with concrete;

— The use of slatted flooring is not envisioned, we remind the reader, because of its excessively high cost and the complexity of handling and storing manure with this system.

Floor covering, therefore, must be selected on the basis of the following constraints:

— Climate—particularly rainfall intensity and frequency;

— Soil type over which housing is built: texture, slope and corresponding drainage capacity;

— The way the housing is used—one can be much less demanding with regard to floor coverings for housing utilized only during the dry season than with year-round housing;

— Animal density: as we have seen, density can vary, with the full open yard, for instance, in proportions of 1 to 3 according to the drainage capacity of the soil.

— The housing system: paradoxically, it is the fully roofed or combined systems where flooring is less of a problem, because here there is protection from the rain.
— The individual sections of the housing: floor covering is a different problem according to which of the three already discussed areas (lying, feeding, exercise), or surrounding areas, it applies to.

Open Yard

It has been repeatedly emphasized that fully open yard housing should only be built on proper, meaning properly-drained, soils. The economic interest of this system lies precisely in the fact that it needs neither roofing nor flooring. The ground may, however, sometimes be treated differently in the three sections of the housing:

— *Rest and exercise areas*: The ground will be left as is or carefully drained so as to increase its resistance to trampling by the animals' hooves. Indeed, in view of the area allotted to each animal, better flooring such as blocking or paving would drive costs up to the point where the system would no longer be competitive with roofed systems.
— *Feeding area*: This area of the housing, a strip 2-3 m wide along the manger, is usually treated the same way as the rest and exercise areas.

Since the animals drop most of their dung in the feeding area, it might be a good idea to treat this area somewhat differently, such as blocking and then flattening it with some mechanical device, for example, or even pouring concrete. In the latter case, the sidewalk system is used: a horizontal platform 2-2.5 m wide is built along the manger. From time to time, the dung and refuse which have accumulated on the sidewalk are cleaned with a tractor rear- or front-mounted dung scraper, taken out of the yard and dropped into a slatted trench where the liquid manure can filter off, resulting in a dry manure which is much easier to handle and transport.

Combined System

The amount of ground area is much smaller in the combined than in the preceding system (1.5-2 m roofed and 2 m open). Here the lying area is under the roof and the exercise and feeding areas comprise a single section in the unroofed portion.

— *Covered area*: As it is not subjected to run-off, there is no point in giving the ground in the lying area any special treatment. The original levelling and tamping of the soil is sufficient, with the proper litter, to ensure the animals' comfort.

— *Open-air section*: This is the feeding and exercise area. This is also the section of the housing where the tractor and trailer cross during manure removal. And lastly, this is the area where most of the dung collects.

If rainwater does not run off this section of the yard, it can become a real cesspool where the animals will have a hard time moving about, especially during the rainy season. So the ground in the open part of the combined system should not be left as is, or simply tamped as in the preceding system: there will have to be some sort of treatment, which might include:

— blocking followed by rolling, where drainage is satisfactory;

— concrete flooring in this section where drainage is poorer; this section can be cleaned with a tractor dung scraper.

Unfortunately, we often see stockmen doing just the opposite; proceeding to pave the sheltered area of combined housing whilst leaving the ground in the open section as is. For the same expenditure, the housing loses much of its effectiveness due to the fact that the concrete is not serving any purpose and the feeding area becomes impassable during the rainy season.

Roofing

The housing is entirely covered by a roof, so it is protected from rain and runoff. However, given the high density of animals/ unit of area, different solutions can be envisioned:

Floor covering depending on the availability of litter (straw, wood shavings, or other absorbant, bio-degradable and inexpensive materials).

When litter is available and/or manure has some value, the housing floor can be treated in the same way as the covered part of the combined system. In other words, simply levelled and then covered with deep litter. Here the need for proper amounts of straw (2 kg/head/day) must be re-emphasized to ensure absorbtion of the moisture from urine and dung and hence the animals' cleanliness and comfort.

Moreover, where the animals are watered with an automatic drinking-bowl, it is a good idea to pave a one-meter wide area all around the bowl. On the other hand, where litter is scarce, a sloping concrete platform will be built, and there will be a semi-liquid manure drainage channel running along the feeding passage. With this system, the housing is automatically cleaned by itself as the animals trample the manure into the channel.

Flooring of annexe areas: The feeding passage has to be used several times a day by workers using a trailer (preferably animal-drawn). So this passage must be passable at all times. The ground will have to be blocked and rolled, cobbled, or, the more economical solution, asphalted.

When the housing is set up on a concrete platform, a rather wide cement channel (at least 30 cm wide and deep) should be installed immediately in front of the manger so as to allow a worker to clean out the dung trampled into the channel by the cattle. This cleaning is done from time to time. The holding and circulation passage is only used by the cattle on rare occasions—at the end of the breeding/fattening cycle and perhaps during periodic weighings. So nothing has to be spent on treating the ground in this passage; it can be left as is.

Manger

The manger fulfills a dual function in the housing systems

described in this report: as a feeding area and as a partition separating the cattle from the outside. Special care must therefore be taken with its construction. Its dual function means that the manger is one component of the housing which receives the most abuse from the animals as they rush onto it when the concentrates are distributed. It is therefore a good idea to overdimension the materials used so that the manger will not be quickly wrecked or damaged.

It should also be noted that the design and size of the manger influence the ease with which the animals absorb the fodder they receive. Independently of waste (fodder which has been refused or trampled upon after falling from the trough), the way the manger is built has a direct influence on the animals' daily consumption and hence on the technical and economic consequences of the operation.

Because of this, no matter what materials are used it is imperative to respect those standard dimensions which have been worked out after lengthy trials. These refer especially to the diameter of the trough, its height above ground and the space between it and the withers bars. By the term "manger" we include the entire feeding-trough apparatus together with its protection devices such as the withers and protection bars which keep the animals out of the trough.

Various materials can be used in the construction of the trough:

- Wood and iron, with which movable troughs can be built;
- Reinforced concrete and parpen;
- Semi-circular irrigation channel pipes where available.

Different Materials of the Trough

Wooden or sheet iron trough

Economic reasons may determine the choice of one of these two materials, though they can hardly ever compete with the other solutions suggested, in order to increase the flexibility of utilization of the building, which may have been remodelled in such a way that the easiest access is by movable troughs, but also for specific technical reasons, since these materials are light enough to allow the installation of a rack-bar to adjust the vertical height of the manger.

- — Wood is the easiest material to work with, but it is not available everywhere at acceptable prices. Another factor is that wood mangers usually have a shorter useful life than do mangers built of other materials.
- — Iron is also a rather easy material to work with, but it needs more equipment to work with (welding equipment) than does wood. And its cost is often excessively high in the developing countries. And iron mangers are often attacked by certain rather acid feedstuffs, such as silage, and by moisture (unless the iron is galvanized, which ups the cost further). Nor is their useful life necessarily longer than that of wooden mangers.

The construction of wooden or sheet-iron mangers is subject to the afore-mentioned building standards. One necessary component is a standard protection device, no matter which material is used. The basic advantage of sheet-iron or wooden mangers is that they are lightweight—elements up to 6-8 meters can be handled easily. This means a rack-bar system can easily be installed so that height can be regulated in terms of the ground level of the housing.

By the time the breeding/feeding cycle is over in a fully-roofed, deep litter, small space housing arrangement, the ground level may have risen over 50 cm. This rack-bar can be made from two metal bars which are slid under the trough into slots or holes in the supporting posts and the ends of the trough.

The use of this trough is particularly recommended for fully roofed loose housing, where the trough is protected from inclement weather and can be raised in height as the litter accumulates. Its use may also sometimes be recommended in combined housing using remodelled buildings where lateral access is impossible. Should this be the case, the trough can simply be taken down to allow dung removal or entrance or exit of the animals.

Parpen or reinforced concrete manger

These materials do not always mean either the most economical or the most flexible solution. In addition to the fact that concrete and breeze-block managers can neither be moved or adjusted in height, they also present the disadvantage of preventing the installation of certain manure removal systems. Still, if breeze-block and reinforced concrete often meet with the approval of livestock raisers, it is because they can usually be used by fairly unskilled labour, and especially because the installation is a very sturdy one which can be amortized over a ten to twenty year period.

Parpen is easily manufactured on the farm with a minimum of equipment and at relatively low cost. However, it is not always resistant enough. Reinforced concrete, a relatively easy material to work, has almost unlimited resistance. But its cost price is often considered excessively high. The parpen or reinforced concrete manger is built by making two parallel partitions 60 cm from one another. Parpen partitions can be 15 cm thick and reinforced concrete 8-10 cm.

The partition on the side of the housing is 60 cm high. The partition on the side of the feeding passage is 75 cm high. In this manner the cattle are unable to push the fodder outside the trough (and outside the yard) with their muzzles. The area between the two partitions is filled with pebbles up to a height of 30 cm, leaving a trough depth of 30 cm on the side where the cattle feed.

After plastering the whole arrangement with a mortar rich in cement, the standard protection device will be installed as per the other manger models. Wherever semi-circular irrigation pipes are unobtainable, the use of this kind of trough is recommended for fully open yard and combined housing systems. As a matter of fact, since this kind of trough is attached to the ground by foundations, the masonry trough can obstruct the passage of dung towards the feeding passage. This would make it difficult to set up a sloping concrete platform, which is of interest in areas where straw is too scarce to enable the deep litter system to be used.

Irrigation pipe trough

This is the recommended material for use when the farm is at reasonable distances from where the pipe is manufactured. These semi-circular pipes of vibrated concrete have an interior diameter of 60-80 cm. Trapezoidal pipes, with an interior diameter of less than 60 cm have proven less resistant and are to be avoided, as they do not offer sufficient capacity for the distribution of voluminous feeds such as fooder (silage, hay, green fodder).

This is the simplest and quickest kind of manger to build. All the pipes are laid onto their prepoured supports at the same time, thus cutting down on labour supervision. It is also a very sturdy material, resistant to the shocks of bad weather as well as the weight of the animals. Its useful life is virtually unlimited. In the final analysis, this is the most economical solution, especially when the pipes can be bought second hand or at bargain prices (should they not be absolutely waterproof, and thus unusable for transporting water, this defect has no effect upon their use for feeding-troughs).

A detailed description of how to build feeding-troughs from irrigation pipes is given in the following pages. Let us just mention at the outset that use of this material is limited only by transport and handling problems (where the pipes are available). The weight of each component, measuring 6-8 m in length, is such that they have to be moved to the building-site by large trailers, and these cannot always get through the trails leading to the building-site.

Unloading should preferably be mechanical, so a crane ought to already be available on the site. In light of the costs involved in moving the pipe, its use would not seem to be recommendable unless the building-site is less than 100-200 kms from where these pipes are available.

The semi-circular irrigation pipe trough can be recommended for use in all types of housing. In any case, to the extent this trough is too heavy to be able to be adjustable in height by means of a rack-bar, its use will preferably be reserved for fully open and combined housing systems. The existence of a space of about 30 cm beneath the trough and the ground means that irrigation pipes can be used with concrete platforms which are so useful in areas where straw is scarce.

Components of the Manger

Though the manager may well be built with various materials, the authors have decided to provide a detailed description of a manager, its protective devices, and assembly, based on the use of irrigation pipes. This does seem to be the material with the best cost/strength ratio in most cases. And it is also a rather original technique, which deserves an explanation. Concerning the protection system, this is identical, whatever material is used in building the manger.

Supports for the irrigation pipes

Various materials can be used to support the irrigation pipes. The manufacturers of these pipes often deliver them with vibrated concrete supports designed to fit the outer sides of the pipes. But when such supports are not available, they must be built from bricks or breeze-blocks. The use of such lightweight materials as wood and metal is not recommended as they are too weak to support the weight of the pipes.

Where the supports are delivered with the pipes, they must be bedded in accordance with the following standards:

— Distance between the axises of the supports must be exactly equal to the length of the pipes used so that where battery housing is used, the end of each pipe can be laid on one-half of the support (the other half is for the next pipe). This material is so strong that no intermediate support is needed.

— Support thickness varies with the material used: about 15 cm is enough for reinforced concrete. A row of bricks where brick is used.

— Support height is basic as the height of the trough is determined by it. To respect the height of the trough, the lowest part in contact with the pipe should be about 20-30 cm off the ground, according to whether the pipe measures 60 or 80 cm, respectively, in diameter, and about 7 cm thick.

— Horizontal supports on sloping ground would not be a desirable element; the essential priority is the height of the edge of the manger with respect to the ground. In the no-roof, fully open yard system, the manager will be set up on the highest part of the site so as to avoid the accumulation of water in the feeding area, which is where most of the excreta accumulates.

Manger proper

As has been emphasized, there are many advantages to the use of irrigation pipes for feeding-troughs. But these pipes must be used in accordance with a number of highly specific recommendations. The pipe is laid asymmetrically on its supports. The two edges of the pipe are not level with respect to the ground. The side of the pipe facing the housing has to be about 15 cm lower than the side facing the feeding passage. It is this asymmetry which prevents the animals from knocking their fodder over the edge of the trough with their muzzles.

Where very voluminous feed is distributed (hay, straw), the asymmetry of the trough can be added to by fastening a plank about 20 cm wide in a vertical position above the outer edge of the trough. The pipe has to be laid at the proper height with respect to the ground: the right height is 60 cm aboveground for the side the animals are on and 75 cm for the side facing the feeding passage. How accessible the feed is determined by the height of the manger on the animals' side of the trough. Obviously, what is needed is an average height applicable to both 100 kg cattle housed immediately after weaning and 450 kg young cattle ready to leave the housing—slaughter for the males and breeding for the females.

It is imperative to respect this standard. Anything higher would block the animals (especially the younger ones) at dewlap height, and would keep them from getting to the bottom of the trough—to the concentrates, for instance. Anything lower would be wrong because the animals grow in size and the ground level rises where deep litter is used. The result would be that the manger would be filthied, the animals would have to kneel to feed, and there might even be accidents as the animals, with their feet in the trough, were jostled by the others.

Stockmen often build their mangers level without taking into account the slope of the site. The result is that the edge of the manger is the wrong height along much of the housing. If a livestock raiser is bothered by having a sloping manger, then he will have to agree to the outlay for terracing prior to construction of the manger.

Support posts for the withers bar

Various materials can be used for making the posts to support the manger protection system (withers and protection bars):

The use of wood makes it considerably easy to fasten on the elements the post is intended to support. And it is usually cheaper. Hardwood sapling poles 15 cm in diameter are preferable to sawed-off pine wood which has to be 20 cm through. In any case, the posts will presumably be treated to prevent rot.

The more aesthetic metal tube, with an external diameter of 80 mm, can also be used. But it is more expensive and is more difficult to attach the withers and protection bars to. The reinforced concrete post may sometimes be the most economical, but, here again, there will be problems with attaching the protection bars to it.

The dimensions of these posts vary according to the bedding system selected, no matter what material is used. The above-ground measurement of the post is that of the fences: 1.60 m. If the posts are bedded with road metal and coarse concrete, they can be sunk to a depth of just 50 cm, making a total length of 2.10 m. However, if, for economic reasons, one should decide to not use cement and use only tamped earth to hold the posts, they should be sunk to a depth of 80-100 cm, or a total length of 2.40-2.60 m.

Bedding the posts: Spacing between two successive posts must not be greater than 2.50 m. The substantial pressure put on the withers bar by the cattle when the feed is distributed would mean that any greater spacing than 2.50 m would make the protection device too fragile.

The posts are so arranged as to touch the top side of the trough on the side of the housing: if the posts were placed away from the trough, the withers bar would keep the cattle too far away from the trough. And lastly, for prolonged use of the installation, it would be desirable to protect the posts from attack by urine and litter, especially where the deep litter system is used. This is particularly important when metal posts are used. Set into a block of concrete about 30 cm above ground level, they will be sufficiently protected.

Withers bar

The withers bar has an essential function. This is a horizontal bar which, as its name indicates, is located at the animals' shoulder height so as to keep them from getting too far into the manager whilst at the same time allowing easy access. With the withers bar, it is at the same time possible to:

- prevent the cattle from getting their feet into the trough;
- force them to feed perpendicular to the trough and therefore reduce to the minimum the trough width occupied by the individual animal;
- keep the animals from getting out of the housing. We might say the manger fulfills the dual function of providing food and fencing the cattle.

Materials suitable four use for withers bars have to be sturdy enough to resist the shoving action of all the animals rushing at once to get to the trough to eat their favourite foods, such as concentrates. The ideal withers bar is the metal pipe with an internal diameter of 50 mm and an external diameter of 60 mm. However, if this is not available, a smooth, cylindrical wooden bar, with a diameter of at least 120 mm, can be used. It is not advisable to use a board. It would be too fragile, considering the distance between the posts.

The bar is fastened on the side away from the housing. In other words, over the manger. Though it would seem that to fasten the bar on the animals' side would make it more solid, this is to be avoided as it would push the animals back from the manger at a distance equalling the sum of the diameters of the post and of the bar.

The bar is fastened in the same way for a movable or stationary manger. A metal bar will be fastened by two forged hasps bolted onto the post. A wooden bar will be fastened by two bolts going through both the post and the bar. For movable mangers, regularly spaced holes will be drilled into the posts, so that the withers bar can be raised at the same time as the manger is moved. The distance between the upper edge of the trough on the side of the animals and the withers bar must be 40 cm. For animals weighing under 200 kg., it can be cut to 35 cm and for animals weighing over 500 kg it should be increased to 45 cm.

Protection bar

The trough and the withers bar would not be enough to keep the young bulls from climbing over the manger without a protection bar above them. In the final analysis, it fulfills the function of being the highest part of the permanent fence. One can, in fact, given the aforementioned measurements, calculate the height of the withers bar at a bit more than 1 m off the ground, which would not be enough to keep the animals from jumping out of the yard.

It is rare that the protection bar has to bear the weight of the cattle. Its part is rather to dissuade them from trying to jump or climb over the manger. The materials used to make it, therefore, are more lightweight. They will be the same as those used in the permanent fencing: 35 mm boards, 100 mm diameter wooden bars or 500 mm diameter metal tubes. They are fastened in the same way; for the pipe, and for the board or wooden bar. The protection bar is fastened so that the height of its upper edge is the same as the top height of the permanent fencing—1.60 mm from the ground.

Drinking-bowl

Young cattle need a permanent, always accessible source of water. Proper watering is an a priori in successful cattle raising. The poor results which have sometimes come out of conventional stanchion stable establishments which are not equipped for frequent watering of the animals have to do with the fact that insufficient watering brings about diminished fodder consumption and therefore a lower performance platform.

Among those factors influencing the daily intake of water are liveweight, the kind of fodder distributed, dry matter content of the cattle's diet, individual performance, climatic conditions, how often the animals are watered, etc. It can be estimated that the average daily water intake of young cattle weighing in at 300 kg is about 30 litres, which varies substantially with climate and diet. A single manger model has been presented for all types of housing and materials used. However, different solutions can be envisioned for the drinking bowl:

When the farm does not have a supply of running water, the best plan is to store enough water in the housing area to supply the animals' needs for over 24 hours. The best solution in this instance is the large-capacity reservoir tank. But unfortunately it is only suitable for fully roofed open housing. However, when the farm does have running water, the reservoir can be a rather small one if it refills automatically. The construction of any type of loose housing which would not have a permanent supply of water accessible at all times to the animals is definitely not recommended. To take the animals out of the enclosure once or twice a day to drink would inevitably reflect unfavourably on their performance.

Reservoir-watering Trough

Supplying water: The reservoir-watering trough will be installed only on farms without running water. Indeed, the much higher installation and operating cost of this system is only justified where water has to be transported by movable tanks from a rather distant source of water.

A unit of more than 20-30 animals presupposes the installation of a reservoir watering-trough which would require the permanent presence on the farm of a tank trailer and tractor. For smaller units, animal traction is recommended.

Description of the watering-trough: A reservoir-type drinking-trough is made up of a tank with a 50-litre capacity per animal and the accessory protection devices. The simplest solution is a semi-circular pipe identical to the one used for the manger. It must be closed at both ends by a cement partition and laid horizontally with both edges at the same distance from the ground. A pipe 60 cm in diameter and 8 metres long has a 1 100 litre capacity and can meet the needs of a twenty-head lot. A pipe with an 80 cm diameter (internal diameter in both cases) can be used for larger lots. For the same 8 metre length it can contain about 2 000 litres, which can water a lot of nearly 40 head.

If these pipes are not available, the tank can be made of breeze-block or formed concrete: the top of the tank should be 60 cm off the ground, the depth 30-40 cm, and the width 60 cm. These masonry tanks have a capacity of 180-240 litres/linear metre. The use of metal tanks is only justifiable economically speaking if the tanks were purchased for some other purpose and are simply being adapted for use as watering-troughs. Whatever material is used in making the tank, it should receive the same protective devices as the trough; i.e., withers and protection bar. One possible arrangement for the tank/trough is to build it between two lots, whereupon it becomes part of the partition separating two adjacent yards. In this case the protection system has to be set up on both sides of the watering-tank.

The tank/trough is very adaptable to no-roof open yard systems, and, where strictly necessary, to combination-type housing. In the latter case, it has to be installed under the roofed section, which limits the length to 3-4 m (and hence the capacity).

The installation of the tank/trough in fully roofed loose housing is not generally recommended. Finally, it should be noted that carrying water by tank trailer weighs heavily on the budget for feeders/breeders: it is always desirable. to examine the possibility of piping water and thus making possible the installation of one of the two above described drinking-bowls.

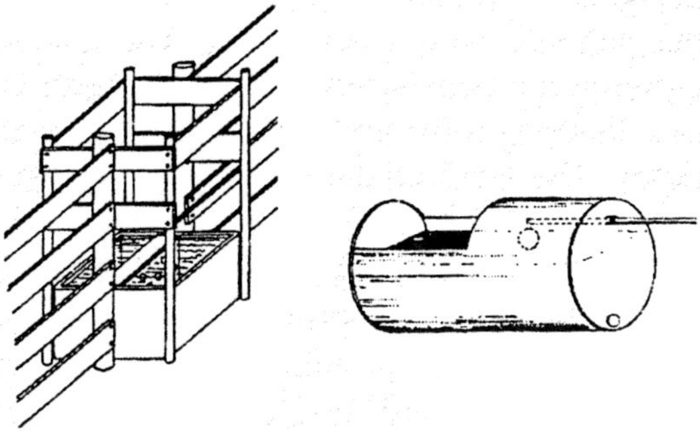

Figure 1.

Constant-level Drinking-bowls

Water supply: Wherever possible, the installation of water under pressure is recommended for young cattle housing:

— plugged into an existing network of pipes; of course the choice of the housing site will have to take into account such a network and its location;

— constructing a water tower with mechanical lifting such as an electric, motor, gasoline or wind-powered pump.

— setting up a reservoir at a higher elevation to provide water for the housing even though refurbished by tank truck, when the spring or well is far away.

It does seem excessively costly to set up a supply of water under pressure, but in actual fact it always translates into considerable savings when compared with watering from tank/troughs, because:

— the automatic supply is a labour-saving device;

— water under pressure removes the risk of a water shortage, which can result in lowered performance;

— with water under pressure, the smaller drinking-bowls, which are better adapted to combination and fully-roofed loose housing, can be utilized.

Description of the watering-tank:

— a 100-200 litre tank can be made of concrete and designed to include protection and withers bars so that the animals cannot get into it. The amount of water entering the tank is controlled by the sort of ball found in a flushing toilet and hidden beneath a sturdy metal cover. The level of the water can thus be maintained constant.

A tank of this type can also be devised much more simply and less expensively from a 200-litre metal tank. A portion of the metal is removed so that the animals can drink from the inside of the tank, and the ball arrangement is protected by the covered portion.

Figure 2.

The use of the constant-level drinking-bowl is recommended for all types of housing at the rate of a single apparatus per lot in the case of a metal tank and one apparatus shared between two lots in the case of a concrete watering-trough. Indeed, the constant-level drinking-trough of local manufacture can usually be an advantageous substitution for the automatic drinking-bowl in most of the developing countries.

Automatic Drinking-bowl

Water supply: The advantages of installing water under pressure in feeder/breeder installations can hardly be overemphasized. The characteristics of the automatic drinking-bowl, which only has a capacity of about 1 to 2 litres, and which refills as the animal drinks, mean that a certain minimum of pressure is needed for correct utilization. So where pressure is low (such as a reservoir of water at not much greater elevation than the housing), this system is to be avoided.

Description of the drinking-bowl:

Usually made of cast-iron and delivered ready for immediate use, it has only to be connected up to the running water supply. It has the advantage of taking up very little room. Unfortunately, it cannot usually be manufactured in the developing countries and must be imported at relatively high cost.

The automatic drinking-bowl is of universal application. It fits all types of housing and is especially suitable for the fully roofed loose housing system because it takes up so little room. One drinking-bowl can accommodate 15-20 head (one at a time, of course), and hence lots housed in combination or fully roofed loose housing.

The location of the drinking-bowl, whether automatic or constant-level systems are used, must conform to the following recommendations:

— *Fully roofed open housing*: The drinking-bowl will be installed in the exercise area, along the stationary side fences.

— *Combination-type housing*: It will be installed just immediately before the roofed area.

— *Completely open yard housing*: The drinking-bowl will be installed in one of the rear corners of the yard.

No matter which type drinking-bowl is selected, it must under no circumstances be installed in the centre of the housing (which would hinder mechanized litter removal), by the movable fences (which would prevent access to farm machinery) and near the manger (hindering access to the feeding-trough).

Actual Housing

Here we wish to discuss all those superstructures which fulfil the functions of protection (roof and protective wall) and partitioning (stationary and movable fences). In choosing the materials for building these elements, the builder ought to demonstrate as much imaginativeness and flexibility as possible.

Indeed, the list of materials given here is by no means exhaustive. And this is the area in which the boldest and most economical solutions can be sought. Common sense and cost should be the only guidelines dictating the choice of materials. In any case, it is helpful to keep in mind the function of each component and consider only those factors connected with fulfilling this function.

— The function of the roof, where there is a roof, is only to protect the lying area from too much sunshine and rain. The kinds of conventional (traditional) roofing already found in most of the developing countries are quite adequate for this purpose.

— The function of the walled-off area, where there is one, is only to protect the lying area from the action of the prevailing winds. It is easy to imagine that this being the case, what is needed is not necessarily a wall but a sort of screen, which can be made from the most diverse materials.

— The function of the stationary fence, as has already been mentioned, is to prevent the animals from getting out of the housing. In this area, each country has its own customs, and these can eventually be used to great advantage.

— The function of the movable fence, is the same as that of the stationary fence. However, it must be able to be opened easily and closed firmly. Experience appears to demonstrate that movable bars are preferable to the conventional stockyard gate.

Figure 3.

Roofing Construction Methods

Roofing construction methods for combined and fully roofed housing are flexible for what concerns the materials to be used. But concerning the actual construction of the roofing, methods must be more strictly adhered to. The materials which can be used are extremely varied. In roofed housing, this is the component with the highest cost per head. So the solution opted for must offer the best cost/strength ratio—meaning the one involving the lowest possible amortization cost.

The frame can be made out of: wood (easy upkeep and relatively resistant), iron (minimal upkeep but expensive), reinforced concrete (very resistant but not reusable should the building be remodelled for some other use). These components can, eventually, be prefabricated. The roof can be made from industrial tile (the disadvantage is that it is heavy and easily broken); from eternit (light and easy to assemble but fragile): from sheet metal (even more lightweight but rusts easily). Conventional (traditional) materials such as puddled clay and thatch can also be used. Though they do occasionally have to be redone, there is no initial outlay for materials and they are excellent insulators from heat.

The construction of the roof must respect a certain number of essential specifications:

The roof frame must be high enough to allow for normal circulation of men and animals, and possibly such farm machinery as the forklift, and it must be high enough after the deep litter has piled up (it can pile up as much as 50 cm). The lowest point of the roof frame must be more than 2.5 metres off the ground. This recommendation is of particular importance in the fully roofed type of open housing, where tractors have to be able to manoeuvre at some point.

Since the depth of the fully roofed housing is 6-8 m and the slope is 20 percent the height of the edge of the shed on the manger side is about 4 metres. The slope of the roof must, in fact, always be oriented towards the rear of the building. There is no problem with the accumulation of water in the circulation and holding passage, since the passage is only used by the cattle once a month.

Figure 4.

The roof must never be oriented towards the front. Because in the combined system, the rainwater would then accumulate in the exercise and feeding area, and in the fully roofed yard it would accumulate in the feeding passage, making them both impassable.

Shelter Wall

Like the roof, the shelter wall is only found in the combination and fully roofed yards, which have to be oriented against the direction of the prevailing winds. In order to protect the animals, the end of the covered part is boarded up (or other materials are used) to provide a wind screen.

The materials which can be used for the shelter wall are also very varied.

The most satisfactory solution is to build a brick, breeze-block or stone wall, but it is also the costliest one. It is cheaper to use sheet-metal, boards, and even thick bales of straw protected by fine wire netting on the side of the animals, even if it has to be reinforced by a stationary fence. Nor ought one to eliminate such traditional solutions as the puddled clay (adobe) wall, which involves no other expense than the cost of labour.

Whichever material is selected, the shelter wall must include a ventilation system and an entrance/exit door.

So as to ensure that the housing will be well-aired, it is a good plan to leave a free space about 50 cm high on the top part of the rear wall and all along the building. This opening, which lets the hot air circulate in the summer, has to be able to be closed off in the winter by windows or even by bales of straw or just by fairly strong sheets of plastic (plastic sacks). We remind the reader that an entrance/exit door allowing access to the housing to both people and cattle has to be built in for each lot. This door is set in the protective wall, and provides direct access to the circulation and holding corridor located behind the building.

The door is set into a corner to facilitate the handling of the animals and should measure 1 m wide by 2 m high. Lastly, in regions with very cold winters, or where there is no prevalent wind direction and it blows first from one corner, then from another, a temporary protection system can be envisaged involving the use of a lightweight curtain (nylon or plastic), stretching all along the shed roof to protect the manger and the ground.

Permanent Fence Construction

Fence construction is one of the basic components of fully open yard housing. It is very limited in the combination system and virtually non-existent in the fully roofed yard system. The permanent fencing is made up of posts, bedded in the ground, and fencing materials. Each of these can be made from different materials, but both must conform to exact building specifications.

The principal materials which can be used:

For the posts, it is usually more economical to use wooden poles with a diameter of about 15 cm, but these are also less sturdy when they have to hold up under long and intensive use. Iron posts 15 cm in diameter and iron pipes 8 cm in diameter are strongest. The actual fence can be made from a series of horizontal bars. These can either be 35 mm thick planks, or metal pipes at least 5 cm in diameter.

Despite the variety of the materials which can be used, certain standard specifications must be observed in building the permanent fence. The height of the fence must be 1.60 m off the ground. This is high enough to dissuade the young cattle from trying to jump over it. There must not be more than 25 cm of free space between the bars. This keeps the animals from sticking their heads through the fence, which might cause accidents and damage to the installations.

The number of bars of which the fence is made up is determined by the total height of the fence, the spacing between the bars, and the width of each bar. For 15 cm boards, 4 bars will be enough; for 50 mm metal pipes, 5 bars will be needed with the bottom bar 35 cm off the ground. How strong the fence is depends on the spacing of the posts. Spacing, of course, depends on the kinds of materials used to make the fence and their dimensions. A 2.5-3 m spacing is usually considered to make the fence strong enough.

The way the fencing is joined to the posts is important. It must be strong, not risk harming the animals, and be easy to repair. The nail and barbed wire are to be avoided. The best solution is the threaded rod, and, when the fence is made entirely of wood, the bolt.

Movable Fence

We remind the reader that all of the three suggested housing types must include a service passage so that the litter accumulating in the yard can be removed from time to time. In fully open yard housing, the service passage is directly behind the manger; in combined housing between the manger and the roofed section; in fully roofed housing beneath the roof. Except during cleaning operations, the passage is closed by movable fences.

The materials used in building the movable fence are the same as those mentioned for the permanent, or stationary fence. Height above ground: 1.60 m. Space between removable bars, 25 cm. Number of bars: 4 or 5 according to their diameter. The only difference lies in the way the movable fence is attached to the two supporting posts. Here there are three possible solutions:

The conventional gate would not be possible, given its four-metre length, unless a lightweight, strong material such as metal pipes mounted on a reinforced axis with a support cable were used. The rigid, movable fence is made up of horizontal bars set on vertical mounts at each end, perhaps supported by one or more slanting braces. The fence is hinged at both ends so that it can open either way, or even be taken off altogether.

Movable bars. Whereas in permanent fencing the bars of the fence are permanently attached to the posts, with the movable fence one has to be able to take the bars away: they can slide freely through plate iron hasps attached to the posts. To make sure the bars are not moved by the animals, they can be held in place by an iron bar running through them from one end to the other.

Care of the Animals

Loose housing is common housing, so when handling is necessary, the animals are handled as a lot. So the handling of young cattle housed in this way is quite different from that of a tied animal, especially in the case of ungelded young bulls. Loose housing therefore needs an additional annexe where the animals can be handled and held (circulation and holding passage and corral). In addition, since the animals cannot be treated on an individual basis in open yard housing, another annexe, the infirmary, should be attached for treating sick or wounded animals.

The handling and holding premises are especially essential for the periodic weighing of the lot. Without such facilities, the weighing process can become an endless, tiresome and sometimes dangerous "corrida". So that the animals will not disperse upon exiting from the yard, they should be channelled into a circulation corridor. The principle of this corridor is a simple one. All that is needed is to build a permanent fence along the back of the housing, and parallel to it. This corridor should not be wider than 70 cm. Each yard gives on to the corridor from the entrance/exit door in the shelter wall in the rear of the yard.

The corridor ends in a corral built of permanent fencing and with a movable fence about 3 m wide to let the cattle scales through. The corral should be big enough to hold the entire lot at once, with space of about 1.50 m for each young bull. When weighing starts, the movable scales is brought into the corral at the end of the holding corridor, the housing door is opened, and the animals are forced to proceed towards the scales. They can neither go back, being pushed by the ones behind, nor turn around (the 70 cm corridor is too narrow), so they have to move forward towards the scales. After the entire lot has been weighed, all the animals are in the corral. They can be brought back into the housing area in the same way, only the scales will not be used this time.

The other lots follow the same procedure.

Handling and Holding

Fully Open Yard

The handling and holding of young cattle housed in open yards is rather exceptional, as it only occurs once a month where the animals submit to growth checks. Partly because of this, the animals are not accustomed to be handled. If the housing does not have the proper facilities, the operation can be dangerous for both the men and the animals. Basically, the design of these facilities must be adapted to the kind of housing, the size of the livestock unit (number of yards) and hence the number of animals. The size and weight of the animals will also be considered.

For small, open-yard livestock involving some tens of animals, the aforementioned system of adding a circulation passage 70 cm wide along the permanent fence opposite the manger side of the building is perfectly suitable. The corral is reduced to a simple enclosure of 1.5 m/head. For larger units, with accomodations for several hundred head, more complete installations should be planned.

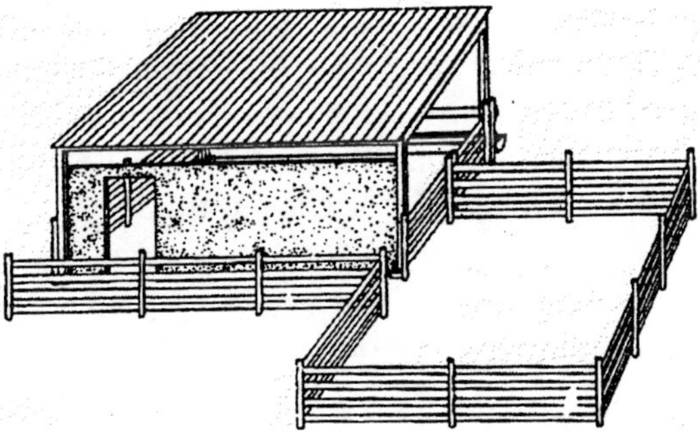

Figure 5.

First of all, out of necessity housing of this type involves batteries of yards built in tandem back-to-back. It might be a good idea to run a service passage 3 m wide between them for vehicles and animals. If this is done, the best thing is to set up a corral at one end of the service passage. It should include at least three sections: a waiting area (1.5 m/head) where the animals wait to be weighed; a holding passage 70 cm wide and about 10 m long, where the animals can be contained and which leads to the scales; a common area where the animals can wait after weighing (1.5 m/head).

Combination and Fully Roofed Systems

For small combination and fully roofed systems holding a few tens of cattle, the system just described can be quite suitable. It boils down to a circulation passage 70 cm wide, narrow enough to double as a holding passage, and which gives onto a simple corral. In this case, the circulation passage is always at the back of the building. The shelter wall is paralleled by a permanent fence. The animals enter the passage by the entrance/exit door already described.

For larger units with accomodations for several hundred head, facilities are of course a bit more complex. A circulation passage for each row of housing and behind the shelter walls will be retained, but there has to be a passage where all these passages lead or meet. This will be a communicating passage at one end of the battery housing, made up partially of movable fences, and giving onto a corral and single holding passage. There really are a large number of design possibilities with regard to corrals and passages. They depend on the kinds of buildings, the number of animals to be handled, and the different operations to be carried out.

Independently of the components just described—the waiting, holding and corral areas—the complex may also include the following components:

- a funnel area as part of the waiting section, making it easier to get the animals into the holding passage;
- a holding or working cage to hold the animals still during such painful operations as foot treatments, tagging, polling, and the like;
- a dip or shower passage for occasional treatment of external parasites;

— a permanent scales; sturdier construction than the mobile models. It would be best to protect this scales by some lightweight construction;

— a sorting arrangement with a multiple access door to facilitate selection of individual animals and constitution of lots;

— a loading platform where the cattle can be loaded and unloaded onto trucks.

There is extensive literature, primarily American, detailing the various kinds of solutions.

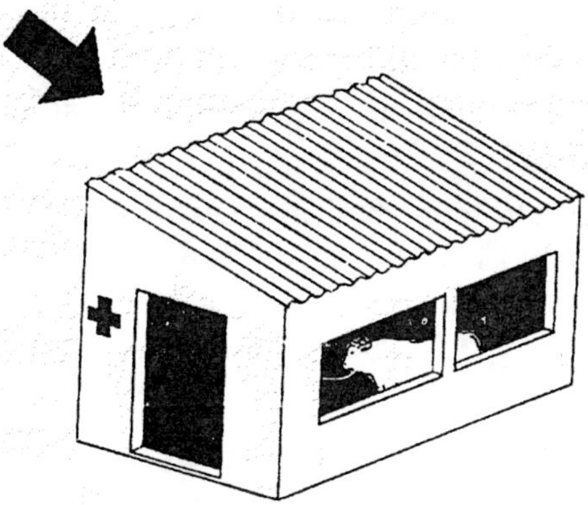

Figure 6.

Infirmary

Concentrations of large numbers of young cattle in the same livestock unit necessarily entails a specific disease situation. To treat the diseases which arise, facilities must be planned as an annexe to open-yard housing, though this is not necessarily the case with stanchion stable housing.

As soon as the first symptoms appear (the animal refuses to eat, stays off by himself, is dominated by the more aggressive of his mates), he must be isolated from the rest in his lot and installed in the infirmary. If a sick animal is left with the lot in loose housing, his recovery may be jeopardized, since:

— he cannot get the proper supervision;

— he cannot get the needed treatment (veterinary care, special diet);

— the healthy animals will keep him from getting his share of the feed.

The infirmary must meet the following specifications:

— it must be so laid out that each sick animal can be isolated, either tied or in his own box where he can be untied;

— it must be so built as to take into account the fact that a sick animal needs better protection from heat, cold, wind and dampness than does a healthy one. So, preferably, the infirmary will be fully roofed even for fully open-yard or combination housing;

— in addition, this will have to be "permanent" construction (something sturdy)—concrete flooring, breeze-block walls, and the like—so that the premises can be periodically disinfected;

— it should be located near the quarters of the workers or herdsmen.

Certain afflictions require permanent surveillance and this may well occur outside of working hours;

— a certain minimum of facilities is required, such as heat, lighting, a store of veterinary medicines, individual drinking-bowl..;

— it must have accomodations for about 3 young cattle out of 100, on the average.

Building Methods

The loose housing models described in this work range from actual construction to simply working the ground. This fact may appear to justify a great deal of variety with regard to installing these facilities. Nonetheless, the actual construction has to follow a strict sequence of operations once the following preliminary work has been completed:

Site selection, which must, we remind the reader, take into account the drainage capacity of the soil as well as the proximity of fodder crops for the cattle (for silage, a distance of more than 3 km between the plot and the silos/housing complex, is definitely not recommended); the availability of water; and the existence of an all-seasons road for shipping farm products and animals off the farm.

Selection of housing system and number of units; this assumes, we remind the reader again, a detailed study of the specific conditions of the kind of livestock specialization involved. Transport onto the site of all the building materials needed before beginning construction. Too often housing construction is halted at the last moment because certain materials are no longer available.

So, a strict sequence of building operations has to be respected, each separate job and special worker following one another in the site so that each operation is carried out under the best possible circumstances. If this sequence is not respected, much time can be wasted. Sometimes, work already completed has to be undone or redone. In any case the work plan and programme chosen must absolutely be respected. Any sudden modification which might disturb the smooth running of the work-site and, ultimately, satisfactory utilization of the housing, is to be shunned.

Site Preparation

The choice of the site for building loose housing for young cattle is extremely important. The length of amortization of this kind of installation can run about ten years. It is recommended that both the present situation and development prospects for the individual farm specialization be taken into account. Particularly:

- the existence of access roads allowing all-seasons transport of farm-produced animals and feed off the farm;
- soil type, which is, we remind the reader, very important in fully open and combination type loose housing;

— proximity of other farm buildings, particularly those for housing the workers in charge of caring for the animals and from which operations will be directed;

— the nature of the feeding/watering system, feed storage and water supply. In effect, feed must of necessity be stored in close proximity to the housing facilities. The daily transport of fodder and water over a distance of several kilometres, often by tractor, adds very heavily onto the cost of breeder/feeder operations.

In calculating the total area of the installations, both the housing proper and the annexes must be considered: feeding passages, circulation passages and open areas facilitating the circulation and manoeuvering of transport equipement. Next, a topographical survey including an assessment of the drainage capacities of the soil. As we have seen, this is one important essential for fully open-yard housing.

The topographical survey is also essential where the land chosen for the site has to be levelled before construction. Lastly, site preparation includes staking-out, which consists of marking off all areas of the building which touch the ground, and which include the supports for the mangers and drinking-troughs, fence posts, roof supports, and the like.

Installation of Supports

The first step in building the housing is to install the support posts for the manger, drinking-trough and roof (if the housing is to be partly or fully roofed).

This is a two-stage process:

— The foundations can be dug by hand, with a tractor-driven drill, or by excavator. This is also when the holes for the fence posts (both movable and permanent) should be dug.

— Foundation depth for manger supports and watering-trough supports has to be calculated so as to allow these building elements to be set at the proper height

So as to ensure that the permanent and movable fences will be strong enough, it is recommended that their support posts be sunk to a depth of about ? the portion aboveground, or roughly 55 cm. The volume and depth of the concrete base to support the roof can vary according to the weight of the roof, and hence of the roofing materials used. With a metal frame with a bearing-surface of about 8 metres and fiber cement roof, the posts can have a volume of roughly one/eighth of a cubic meter and a depth of 50 cm.

A cement-rich concrete is needed for the bedding of the support posts so they will be able to bear the weight of the semi-circular pipes, which is considerable and which is increased by the weight of the water it contains in the case of the reservoir-drinking troughs. Spacing between the manger supports must, we remind the reader, be calculated in line with the length of the pipes used to build the manger. Care will be taken with the drinking-trough supports to ensure that the two posts which keep the pipe in a horizontal position are properly placed. The roof supports usually have a metal bolt to which the frame is bolted.

Building the Roof

Let us recall that different possibilities have been considered concerning the materials which can be utilized for mangers and watering-troughs. Just where installation of the manger fits in the building sequence has to do with the materials chosen for its construction.

Wooden or iron mangers will be installed at the end of the sequence, after the fences have been installed. Concrete or parpen mangers, do not require prior bedding of supports, and so they can be built at this stage. For mangers and drinking-troughs made from semi-circular irrigation pipes: they cannot possibly be laid manually, given the fact that an irrigation pipe 80 cm in diameter and 80 cm long weighs 2 tons. If unloading and laying of such pipes were attempted without the help of a crane, either the supports or the pipes themselves might very well crack.

For this reason, on the day the pipes are laid and as soon as the support base is perfectly dry, a hydraulic crane just like the ones used to load the pipes onto delivery trucks at the factory must be brought onto the farm. By planning ahead in this fashion, all the pipes to be used in the housing complex. (troughs and mangers), can be laid at the same time.

Building the roof, should there be one (fully roofed and combination systems). At this stage the frame is erected and the roof put on. For instance, where metal frames are used, quite a bit of room is needed to install them. Were this left until a later stage, it might be quite difficult for the workers to maneuver among the posts and fences. These should be mounted later.

Bedding the Fence Posts.

We have already indicated that the foundations for the support posts for the withers bar along the manger and reservoir-trough and for the permanent and movable fences around the housing must be dug at the very onset of the work.. In fact, once the pipes have been laid and the frame built, such earthworks could only be carried out under poor conditions.

On the other hand, were the posts to be set too early, they would disturb the progress of the work and might well get broken, especially when the troughs and mangers were laid. Not only this, but since the recommendation is to bed the manger posts so as to lay right next to the upper edge of the manger, bedding could not be carried out satisfactorily until after the pipes were laid.

Special attention must be paid to bedding the posts. The overall strength of the rent of the housing superstructure depends on how solidly these posts are laid in. Bedding will therefore be reinforced (volume of concrete, amount of cement used in the concrete) for those posts subjected to special stress by the animals.

Basically, these posts are intended to support the withers bars of the manger which the animals always rush against as soon as their favourite feed, concentrates, is distributed. The same is true for the posts to support the movable fences, especially the ones barring the service passage temporarily. Not only is this an unusually wide span (4 meters instead of

2.5-3), the fact that the movable fence of course has to be hinged instead of bolted to the post makes the post particularly vulnerable. Lastly, the posts of the holding and circulation passages also have to be bedded very solidly.

Fencing

Only when all the building elements which have to be set into the ground have been installed (manger supports, roof frame, fence posts) can the flooring work be initiated: paving or asphalting the feeding passage, blocking or cementing the outdoor area of combination housing—as needed. It should be noted that whenever the water for the drinking-bowls can be piped in from a network of running water, the pipes should be sunk beforehand so that they cannot be damaged by animals or machinery.

Obviously, were the flooring laid before this stage, it might be necessary to futilely destroy part of the work done in raising the supports or bedding the fence posts. On the other hand, one thing which must absolutely be avoided is to lay the floor after the permanent fences and shelter wall have been built. They would so hinder the progress of the work as to have to be temporarily taken down.

Raising the roof is the very last stage in the process, together with the installation of the withers and protection bars before the manger, the movable fences at either end of the service passage and permanent fences as planned. At the same time, the shelter wall partition and the entrance/exit door at the covered end of roofed or semi-roofed housing will be erected.

It happens too frequently that the building of the circulation passage, upon which satisfactory handling of the animals depends, is left until some later stage. This is a mistake which must be avoided. It is really inconceivable to think of receiving the young cattle before both the housing and its annexes have been completely finished and the resting area strewn with a few bales of hay, ready for occupancy.

REFERENCES

Ruoho, O., "Elimination of Salmonella infection in a beef herd in a loose housing system", *Suomen Elainlaakarilehti,* 1996.

Sonnenberg, H., "Straw comfort: mechanical straw preparation for loose litter livestock husbandry", *Landtechnik,* 1998.

Suzuki, K, Oizumi, C, Kobayasi, M, Mori, T, Wakamatu, M, Sajiki, S., "Effect of the wind blowing right below down by the fan for the beef cattle house", *Bulletin of the Chiba Prefectural Livestock Experiment Station,* 1999.

4

Sustainable Pastures Management

Pasture is at the heart of organic livestock management. According to the National Organic Rule, all ruminants must have access to pasture... and, pasture must be managed organically using strategies that improve soil and water while increasing the nutrient value of the pasture. Controlled, rotational, or management-intensive grazing has increased forage production for many producers. Skilfully using livestock to harvest forages leads to improved soil fertility, a diverse, dense, and useful pasture ecology, and an extended grazing season. Fertile soil and productive pastures, in turn, support healthy animals.

Well-managed forage systems contribute to an operation's sustainability in several important ways:

— Lands most susceptible to erosion can be maintained as permanent sod.

— Land used for row crops benefits from a year or more in pasture as part of a crop rotation plan. The life cycles of annual weeds and other crop pests are interrupted during the pasture years of the rotation.

— Soil health improves as the content of organic matter increases under good grazing management.

— Soil structure improves over time as compaction and hardpan is reduced.

— RuminanPasture is at the heart of organic livestock management. According to the National Organic Rule, all ruminants must have access to pasture... and, pasture must be managed organically using strategies that improve soil and water while increasing the nutrient value of the pastures (cattle, sheep, deer, goats) thrive in a better balanced agro-ecosystem and produce milk, meat, and fiber from grasses that cannot be digested by humans. Livestock eat excess plant materials while animal wastes contribute nutrients for plant growth.

— Marketing meat, milk, fiber, and other animal products can diversify producer income.

In the not-too-distant past, farmers more fully integrated crop and livestock enterprises as a matter of course. Grain produced in field rotation was either sold or fed to livestock, depending on market conditions. Cropland was rotationally seeded to forages, usually for several years. Land not suitable for crop production was grazed. Animals also foraged after-harvest crop residues and the remains of failed crops. These time-honoured strategies are not totally absent from today's agricultural landscape; however, a better integration of crop and livestock enterprises is a necessary step toward the goal of sustainable pasture lands.

Planning and Goal-Setting

An excellent goal is to produce enough good-quality forage to sustain livestock over as much of the year as possible. Then choose the livestock that can best use it. Of course, different livestock species and classes of livestock have different feed requirements and forage preferences. Most cow-calf operations, for instance, have lower forage nutrition and soil fertility requirements than do most dairies. Consistent production of high-quality forage under current management makes a dairy or stocker enterprise an option to consider.

Otherwise, a different class of cattle, sheep, or other ruminant may be more suitable to your specific site and management capability. In setting production goals for any livestock enterprise, consider the economic return per acre rather than production per animal. This is a change from traditional thinking. Compare pounds produced per acre or per dollar invested rather than weaning weights or shipping weights.

Renovating Pastures and Establishing New Ones

Planting a new pasture offers the opportunity to choose forage species and varieties suited to the livestock type adapted to the soil and climate. Efficiency is further enhanced by matching the season of maximum forage production to the period when livestock can best use it or most need it. Further, planting a diverse mixture of forages with differing maturities provides a high-quality, longer grazing season.

Many pasture problems—such as sparse plant cover, weed invasion, and slow growth—are caused by poor grazing management. If this is the case, establishing a new pasture will not solve the problem. Newman Turner, in Fertility Pastures and Cover Crops, observes that good grazing management can transform poor grazing land into healthy, productive pasture. On the other hand, newly re-seeded pastures quickly become poor again under bad management. Thus, a careful assessment of management practices is usually the best place to begin to make forage systems more profitable.

Grazing System

Many managers use controlled grazing plans instead of continuous grazing to increase forage utilisation and profits. In a system of controlled rotations, pastures are subdivided into paddocks—fenced acreage of any given size. Livestock is moved between paddocks at frequent intervals, giving animals access to a limited pasture area over a short period of time. The animals do not return to a paddock until the plants have recovered and regrown to the desired height for grazing (usually six to eight inches). As a result, the plants have time to recover, the roots maintain energy reserves, and the livestock always have high quality forage.

Knowledge of forage plants and animal-pasture interaction is necessary to the success of this type of grazing plan—and frequent attention to both is essential. This is why these programmes are often referred to as "management-intensive" grazing systems. Controlled, intensive, and rotational grazing are other terms loosely used for this type of grazing management. Rotations can vary from once every couple of weeks to every 12 hours. Decisions about when to move livestock are based on the seasonal amount of forage available, the rate of forage growth, and the number and type of animals grazing the paddock.

The number and size of paddocks is also considered. Typically, grazing animals are moved quickly through paddocks during periods of rapid plant growth. In the fall, quick rotations keep grasses from going to seed and preserve forage quality. This strategy can delay for several weeks harvesting of forage as hay, allowing for hay to be put up during a dryer time of the season. During other seasons, the grazed area is usually rested long enough for plants to replace carbohydrate reserves and to regrow. A primary strategy of controlled grazing is to use fencing and livestock movement as tools to manage forage growth and protect it from overgrazing. If managed well, these systems produce more forage and the animals always have access to tender, high-quality vegetation that results from controlled grazing.

Profit is the difference between the cost of production and the price received for a product. Most producers do not control the price they will receive for their livestock. Lowering the cost of production is a clear means to increase profit. Costs go down as less feed is purchased and as animal health improves. The key to profitability is to emphasize a decrease in per-unit costs of production over a simple increase in production.

Many grazing managers—or graziers—claim that controlled rotational grazing improves pastures and the profits to be made from them. Forage utilisation improves even under high stocking rates when the animals are moved at the right times. Livestock health improves because forage quality and quantity is better. Soil fertility is better because most nutrients cycle through the animals and remain in the paddocks. This can also reduce the need for purchased fertilizer.

Plant Community Changes

In a continuous-grazing system where animals are given free choice, they will eliminate the most nutritious or palatable plant species, because they graze them repeatedly. Root reserves of these preferred species are eventually exhausted, and the plants die out. Fescue, bermudagrass, and white clover persist under continuous grazing because their growing points remain, even when the plants are grazed heavily. In a controlled-grazing system, animals don't have access to all the plants in the pasture at one time.

Plants are allowed sufficient time to re-grow and restore their root reserves. Eventually, the plant community becomes more diverse under this type of grazing system. There is less competition for the same soil minerals, and plants thrive in the specific microclimates where they are best adapted. Producers report that native grass species and many legumes spontaneously appear in their pastures as rotational grazing systems are adopted. In fact, many advise new graziers to "plant only fence posts" in the first three years of intensive rotational grazing, because the plant ecology will change—for the better.

After three years of controlled rotational grazing, analyse the results. Should new forages be added to the pasture to meet specific production or management goals? If a goal is to extend the grazing season to reduce feed costs, new species might be added to existing pastures. Special-use paddocks might also be considered. For example, in southern pastures with cool-season grasses, the summer slump is a time of low forage production and potential health problems, especially from endophyte-infected fescue.

Native grasses or plantings of summer annuals can fill this gap in the grazing season. In the Midwest, the grazing season may be extended into the winter by strip grazing crop residue. Stockpiled fescue or other grasses, if carefully rationed, can support several extra months of winter grazing, even where there is some snow cover. Small grains offer options for fall, winter, and/or spring grazing, depending on regional climate conditions.

Good Fertility Management

Grazed pastures need less fertilizer than those that are hayed. Animals actually use up very few of the nutrients from the plants they eat. Most minerals are returned in animal wastes as part of a natural cycling of nutrients. Phosphorus is excreted primarily in manure, and nitrogen and potassium return in urine and manure. As long as wastes are evenly distributed throughout the grazing area and biological agents such as earthworms, dung beetles, and soil bacteria are active, the system should be relatively stable.

Good fertility management includes a regular walk through the paddocks to monitor pasture production and to see where specific grasses and legumes thrive. Notice that certain plants tend to thrive under certain soil moisture and fertility conditions. The types and locations of weeds can also indicate how a fertility programme is working and help identify special situations such as wet areas.

Various plants contribute to soil fertility. Legumes increase the total nitrogen content of the soil. Deeply rooted plants such as alfalfa, warm-season grasses, trees, and some weeds bring up other nutrients from deep in the subsoil. These nutrients remain in the top layers of the soil when the vegetation decays and then become available to other plants nearby.

Periodic soil tests and forage analyses are tools to monitor a pasture's status. Soil test results indicate the levels of mineral nutrients in the soil. Forage analysis is a way to test whether nutrients present in the soil are actually being used by the plants. Many Extension offices offer forage analyses; when requesting this service be sure to specify whether test results will be used to balance a feed ration or for soil fertility decisions. Independent laboratories are available if your local

Extension doesn't offer this service. The ATTRA publication Alternative Soil Testing Laboratories is available online or upon request.

Soil test results include fertilizer recommendations based on information the farmer provides about field history and planned use. Remember that these recommendations can vary depending on assumptions by the lab. For example, a recommendation may not be entirely accurate to produce grazing forage if the lab doesn't take into account recycled nutrients by the grazing animals. Lab fertilisation recommendations may be over- or underestimated, depending on whether forage is harvested and removed or grazed on site. Use common sense to interpret soil tests, but keep them to monitor changes in soil chemistry and nutrient levels.

A special test to determine micronutrient levels may have to be requested. It is good to check these levels, since they can be critical to soil—and animal—health. When soils show deficiencies in essential micronutrients, supplement either the animals and/or the soil. Soil organic matter (SOM) is monitored to determine the general health of the soil and its biological residents. You may have to request and pay extra to include SOM in your soil test.

On the soil test report, SOM includes any living or partially decomposed materials, as well as humus, the final product of biological activity. When SOM is relatively high, it contributes nitrogen and helps make other mineral nutrients more available to plants. Adding composted animal manure is one way to increase SOM. Likewise, leaving a thin layer of organic residue on the soil surface contributes to SOM, and it shades the soil and feeds the soil organisms. Some simple methods to assess soil characteristics require just a shovel and a few other widely available pieces of equipment. The ATTRA publication Assessing the Pasture Soil Resource describes several tests that can be used periodically for a quick assessment of the soil.

Soil Amendments

Carefully consider whether purchased amendments are economically justified. If soils are the limiting factor, buying inputs to improve the soil is a wise, long-term investment. In such cases, improvement in soil fertility is key to building a dense, lush, and healthy pasture. Such pasture provides good nutrition to grazing animals, and wastes contribute to further build the productivity of the land. A simple pH adjustment can increase mineral availability in most soils. Legume growth in mixed pastures that tend toward acidity will benefit, and in turn increase available nitrogen and add more organic matter to the soil. Lime is used to raise the pH, but also is an important source of calcium. It is also less expensive than many other purchased fertilizers. The ratio of calcium to magnesium and potassium is important in itself.

Composted animal manure might also be an excellent investment because it adds fertility and benefits soil microbes. However, if manure is applied to the same pastures over many years, phosphorus can build up. Excessive phosphorus levels in soils and the threat of phosphorus-saturated soils leaching soluble phosphorus are serious concerns in some parts of the country.

One situation where fertilizer purchases are often appropriate is in grass dairy operations. Because grass dairies compete with grain-fed systems, producers must provide continuous access to the highest feed value forage available. Likewise, grass-finished meat animals should have plenty of high-quality pasture to gain weight quickly and consistently during the finishing period. Both of these enterprises have potential for good profitability when well managed. Nevertheless, fertilizer inputs are justified only if existing pastures are under full use. The important point is to base decisions on an analysis that compares input costs to the profits or overall benefits that might be generated. Another excellent resource to understand fertility in grazing systems is Nutrient Cycling in Forage Systems, the proceedings of a 1996 conference in Missouri.

Organic Matter

Some recent research has focused on the many organisms that make up a healthy soil ecosystem. Plant root systems work together with tiny plants and animals underground in a complex, highly organised system very similar to the one above ground. The soil biological community includes large population of many species of bacteria, fungi, nematodes, mites, and other microscopic animals. Balances among the population are maintained by variations in the amount of food available for each part of the system. Elaine Ingham, Ph.D, a soil microbiologist, has named this system the Soil Foodweb.

Ingham offers a service to test soils for the presence of various organisms. However, she says a grazier can monitor pasture soil health just by testing for soil organic matter (SOM) content, which includes carbon contained in living organisms, fresh plant and animal residues, and soil humus. This type of test measures the percentage of soil that is SOM. Because organic matter levels are harder to maintain in warmer, more humid climates, what constitutes a "high" or "low" percentage varies in different parts of the country. Local Extension personnel or soil scientists can help define these relative values. A single test establishes a beginning point, and subsequent tests show whether soil organic matter is increasing.

Legumes in the Pasture

Legumes increase soil fertility, improve overall feed value of available forage, and extend the grazing season. Bacteria that live in nodules on the legume roots convert nitrogen in the air to a form the plant can use. After the nodules separate from the roots or the plant dies, this nitrogen is available to nearby plants. Even during the growing season, dead leaves fall to the ground and provide extra nitrogen to the pasture system. Compared to grasses, legumes have higher digestibility and higher mineral and protein content.

When introducing legumes into an established grass pasture, first be sure that magnesium and potassium levels are suitable. Then graze the area heavily to set it back. Many producers use a sod-seeder or other no-till seed drill, but some have had luck with frost seeding. This is the practice of broadcast seeding in very early spring into areas where the ground alternately thaws and freezes. Timing must be good to take advantage of these temperature swings. These are conventional practices, and information is widely available about them. For legumes to prosper in a pasture, the grass must be kept short enough that they are not shaded out.

Nitrogen fertilizer favours the grass, and you can inadvertently reduce the percentage of legumes in the pasture mix by adding it. Each species of legume thrives in a particular pH range, but maintaining it between six and seven favours most legumes. Some legumes, such as lespedeza, tolerate more acid conditions. Many annual clovers produce hard seed and will persist in a pasture if allowed to go to seed periodically. Annual legumes that do not produce hard seed must be managed to allow some plants to go to seed every year to keep them in the forage mix. Beyond this, providing for the nutritional and light needs of legumes, along with adequate rest after harvest, should ensure their persistence.

If the legume is established and maintained at about a third of the total pasture, the plants won't need additional nitrogen fertilisation. Research at Michigan State University shows that different combinations of four cool-season grasses with three clover species produce, on average, 14 percent more forage than the same grasses grown alone and fertilized with 200 pounds per acre of nitrogen.

Managing Weeds Strategy

In a controlled-grazing system, livestock can help control tall weeds that re-seed themselves. Because animals have access to a limited area for only a short period, they often become less selective in their grazing. They tend to eat the same weeds—in young, tender growth stages—that they reject as the weeds mature. Many weeds provide good nutrition during this period of palatability.

Mowing before weeds flower and produce seed also helps to control them, although the cost is higher. Another weed management strategy is to graze different kinds of livestock together. Sheep will complement grass-eating cattle in the pasture by consuming broadleaves, blossoms, and seeds, while goats prefer brushy vegetation high in cellulose. A growing number of beneficial insects is becoming commercially available to control thistles and some other perennial weeds. These weed-eating insects are especially adapted to a perennial pasture where habitat is not destroyed or disturbed by annual cultivation.

If local sources are unable to help, ATTRA has information about biological management tools and where to get them. Tall perennial weeds that livestock do not eat can be controlled with the judicious application of a broad-spectrum herbicide, such as Round-Up®. Hand-held sprayers will work, but a wick-type applicator places the chemical on the targeted weed foliage only.

Hand-held wicks are available as well as equipment designed to be pulled behind a tractor or four-wheeler. Also on the market are backpack flaming devices that actually burn the weeds and provide a nontoxic option to control difficult weeds. ATTRA publications Flame Weeding for Agronomic Crops and Flame Weeding for Vegetable Crops provide more detail about this option.

Conserved Forages and Grazing

Providing good-quality forage throughout the year saves considerably on feed costs. Year-round grazing is possible in some parts of the country and is a realistic goal in some regions. Many producers, even those in cold climates, report favourable experiences with attempts to "outwinter" their livestock. Adequate feed and shelter from wind and moisture are critical. Reports indicate that, under favourable conditions, animals seem to prefer being outside where they can forage at will. A sustainable pasture plan should be based on animals harvesting quality forage for themselves as much as possible. Nevertheless, when spring pastures produce more than livestock can use, machine harvest is one strategy to ensure quality forage later in the grazing season.

Allan Nation, editor of The Stockman Grass Farmer, is fond of questioning the economics of owning "heavy metal." It is expensive to maintain equipment and to harvest forage for hay or silage, so it is sometimes more economical to buy hay or hire a custom baler. However, it can be difficult to find someone to custom harvest and process spring growth at the optimal time.

Another challenge to a spring hay harvest is the weather. A spell of good haying weather, if it comes at all, rarely arrives at the perfect time. One option in wet conditions is to harvest, pack, and seal the excess spring grass in bunkers for fermentation. Livestock, controlled by a single wire of electric fencing, can then have direct access to the silage bunkers. Some producers advocate baling high-moisture hay and wrapping it so that it will ferment. Baleage, as the product is called, is a high-quality feed when properly harvested and protected from air spoilage. This is one way to harvest on time in wet springs. However, specialised equipment is expensive for one producer to own and operate, and rental may not be available.

Several producers in an area with similar needs might recover some costs through contractual arrangements among themselves. The amount of plastic used to seal cut forage is a concern for many farmers as well, since it must be disposed of after use.

Irrigated Pasture Systems in the Western U.S.

Many regions in the western United States, including intermountain valleys of the Rocky Mountains, the prairies of the northern Great Plains, and certain arid regions of the desert Southwest, experience short grazing seasons due to high elevation, limited moisture, or a combination of both. Livestock producers in these regions find it particularly important to manage forage and pasture in the most efficient way possible. By integrating irrigated pasture with dryland pasture, range, and hay aftermath, the grazing season can be lengthened and livestock provided with high yields of quality forage.

ESSENTIALS

Conventional wisdom holds that one acre of irrigated pasture in most intermountain valleys provides enough forage for twelve cow-calf pairs for one month. But unproductive irrigated pastures are more the norm, and few producers maintain pasture to its full potential. Productive irrigated pastures are usually the result of successful management of several production factors, including:

- — fertility
- — irrigation
- — species selection
- — grazing management

These factors can be managed.

Soil Fertility

Attention to soil fertility is critically important in irrigated pastures. Pasture establishment is a key time to ensure soil is adequately fertile for the selected forage species to become established and remain productive. During secondary tillage, rock minerals, composted manure, or commercial fertilizers can be incorporated into the soil. In the intermountain regions, it is important to ensure adequate phosphorus and potassium before planting, but nitrogen should be applied early the second spring. Cool, dry springs are difficult on grass seedlings, and nitrogen applied at this time may be appropriated by weeds.

Apply nitrogen only after the grass stand is successfully established. If the stand has a legume component, limit the use of synthetic nitrogen fertilizers. In general, nitrogen fertilization favours grass growth, and phosphorus fertilization favours legumes. Yearly applications of 20 to 50 pounds per acre of phosphorus can significantly increase alfalfa yields and stand persistence in areas deficient in phosphorus. Soil tests are fairly reliable to gauge phosphorus needs, but again, modern soil testing assumes the forage will be harvested and fed on site. Don't underestimate the utility of the mineral fraction of nutrients in the soil, and the natural nutrient cycle that supports pasture ecology.

Whereas most soil nutrients are cycled back to the soil in a grazing system, some nutrients do leave the pasture system in the form of meat and milk. More information on fertility and nutrient cycling can be found in the ATTRA publication A Brief Overview of Nutrient Cycling in Pastures. Irrigation can also have an effect on nutrient cycling. Coarse, porous soils do not retain water as readily as heavier soils, and heavy irrigation can leach nutrients into the groundwater. If the pasture has any slope to it, nutrients can leave in runoff. Ditches, dikes, and proper irrigation scheduling can alleviate this problem.

Grass-legume mixes provide good pasture productivity and animal nutrition and aid nutrient cycling and pasture fertility. Pastures with a heavy clover component can produce up to 200 pounds of nitrogen per acre per year, and can supply 6 to 12 percent of the nitrogen needs of companion grass plants during the growing year. Given these prospects, a producer can optimize the use of soluble and organic soil nutrients by relying on plant species diversity and nutrient cycling from manure, urine, and plant senescence to supply a large portion of pasture soil fertility. More detailed information on this subject can be found in the sections Managing Fertility and Organic Matter.

Irrigation Pastures

Efficient water use is crucial for sustainable irrigated pasture management. Irrigated pastures require about 24 inches of water per growing season. What is not supplied by precipitation needs to be made up with efficient irrigation. Grasses and legumes require about 0.20 and 0.25 inches of water per day respectively throughout the growing season.

So, frequency of irrigation depends on soil texture and, in turn, on water holding capacity of the soil.

Heavier (clay) soils hold more water, up to 2.5 inches per foot of rooting depth, and coarser (sandy) soils hold less water, around 0.75 inches per foot. Pastures have an effective moisture depletion allowance of about 65 percent, which means plants begin to suffer stress after 65 percent of the soil's water-holding capacity has been depleted. For example, pasture soil with a water holding capacity of 1.5 inches per foot, and a rooting depth of four feet, can hold a total of six inches of water. At a 65 percent depletion allowance, 3.9 inches remains available to the plants. If the plants use 0.25 inches per day, an irrigation event that saturates the soil will last about 15 days.

Understanding the basics of soil-water dynamics helps producers make decisions on when to irrigate, especially in areas where water is scarce or energy costs for pumping are high. The Agrimet system is an excellent resource for producers making irrigation scheduling decisions. In addition, the Natural Resource Conservation Service (USDA-NRCS) district offices have access to each county's soil information and can assist producers to determine the water holding capacity of soil types on area farms.

The Irrigator's Pocket Guide, developed by NCAT for the NRCS, is an excellent resource with timely information on irrigation scheduling, system capacity, and general water management. The Pocket Guide has useful information for most areas. It can be ordered from ATTRA by calling 800-346-9140. Always remember to irrigate a pasture immediately after the livestock have been moved, and never irrigate and graze at the same time. Hoof action on wet soil can destroy its structure, resulting in compaction and decreased soil productivity for years to come.

Species Selection

The importance of choosing the right plants to use in an irrigated pasture cannot be overstated. The high cost of irrigation, including initial equipment purchase, energy, and maintenance demand that a producer select the most productive plant species for the region. In some situations, short season problems and low yields can be addressed though proper species selection. Choose long-lived, winter-hardy forage plants adapted to your specific soil type. Plants should be capable of high yields and have the genetic potential to withstand grazing and regrow quickly.

Species diversity is also important, as was discussed in detail earlier. Greater productivity and increased biodiversity are fostered through grass-legume mixes. A grass component in a legume pasture can also minimize health problems associated with bloat. Some non-bloating legume species include cicer milkvetch, sainfoin, and birdsfoot trefoil. For the intermountain West, a mixture of two grasses and one legume provide as many, or more, benefits to pasture productivity as do more diverse pastures in higher rainfall areas.

Choose the right species for the mix, however, because species that mature at different times can result in low quality forage. Creeping foxtail and timothy are both excellent irrigated pasture grasses, but foxtail matures several weeks before timothy. Red clovers and vetches usually do not persist as well as alsike clover, white clover, and alfalfa in the intermountain regions. Some good substitutes for alfalfa in irrigated pastures are sainfoin and birdsfoot trefoil, which, unlike alfalfa, are tolerant of high water tables. A very common seed mix for irrigated pastures in the intermountain West is meadow brome, orchardgrass, and alfalfa.

Warm-season grasses are sometimes a good choice for the Southwest and Great Plains, and can result in substantial livestock gains and milk production when managed intensively. Warm-season annuals such as sorghum and sudangrass are good choices for rotational or strip grazing, and are very good if the pasture is used in a crop rotation. Cool-season grasses such as brome, ryegrasses, timothy, and cereals are often higher in digestibility and crude protein, and are more adapted to intermountain, inland Pacific Northwest, and Great Plains regions.

Check with your local county Extension office or conservation district for recommendations on forage species particular to your area. For general purposes, please refer to the Alberta Forage Manual and the Intermountain Planting Guide cited at the end of this publication. These two guides are excellent sources of information for anyone growing pastures and forages in the intermountain West or northern Great Plains. A list of forage species for Montana and Wyoming—widely adapted to irrigated pastures in many western states—is enclosed.

Forage Cropping Systems

Many western ranchers grow alfalfa hay to provide high quality feed to late-gestation and calving cows in the winter. Most alfalfa fields remain productive for six to eight years in the intermountain West. As sward density diminishes, the stand is generally terminated and placed into small grains for a year or two. This rotation has its benefits. Tillage and crop differentiation allows the producer to break the pest cycle. And termination of an alfalfa field offers an opportunity to augment ranch forage assets with quality pasture while extending the grazing season as well.

For example, a producer might terminate the alfalfa and plant winter wheat in the fall, and then overseed the field with annual ryegrass in the spring. The wheat can be taken as grain, silage, or hay in the summer, allowing the ryegrass to grow for late summer and fall grazing. The same can be done with spring-planted barley. The result of this cropping system is a high quality pasture that can be intensively managed with high stocking rates, thereby resting native pastures that might otherwise be grazed the same time each year. Other systems that work well to extend the grazing season:

— Stockpiling perennial grass or legume forage for fall grazing.

— Early season grazing of winter wheat and subsequent grain harvest.

— Planting perennial grass pastures for use as winter

standing forage, e.g., Altai wildrye, which maintains quality well when dormant and stands up under a snow load.

Grazing Management

Complementary grazing is a system in which livestock are grazed in annual or perennial seeded pastures in the spring and fall, and are taken to native range in the summer when the native grasses are in their prime. This system uses each pasture when it is at its peak in quality and quantity, and it is commonly used in western states to supplement range and extend the grazing season. Within this context, western producers are familiar with continuous grazing.

The size and scope of grazing units, coupled with the use of public grazing allotments, often preclude fencing and other necessary infrastructure to support intensively managed rotaional grazing. In addition, most producers who graze irrigated meadows also hay them once or twice during the growing season, and only graze them for hay aftermath. For this reason, irrigated meadows tend not to be managed intensively for grazing, as they are seen to be more valuable for winter feed than for summer grazing. After all, that is what the mountain meadows are for.

However, for the producer who wishes to scale back on hay production, the irrigated meadows can be used for grazing during the growing season, and upland meadows that consist of bunch grasses like Altai wildrye can be stockpiled for winter feed. Altai wildrye typically remains a high quality forage well into the dormant season, and large bunch grass type holds up well under a snowload.

Producers who choose to develop a rotational grazing system on their irrigated meadows can realise better animal gains per acre and reduced feed costs associated with feeding the cow herd in the winter. The ATTRA publication Rotational Grazing for a general introduction to this type of grazing system. For most cool-season bunchgrass species, 18 to 27 days rest is adequate for substantial regrowth without allowing the plants to become too mature. A problem that can occur in short-season regions is forage maturing in the last pastures to be grazed before the livestock get to it. To deal adequately with this situation a producer might turn livestock in to the first pasture early, maintain a quick rotation, and then slow it down as the season progresses.

A good formula to estimate an initial pasture stocking rate is:

$$\frac{Pasture\ size\ (ac)\ x\ pasture\ yeild\ (Ib\ /\ ac)}{0.036\ x\ avg.\ animals\ wt\ (Ibs)\ x\ grazing\ season\ (days)}$$

For example, assume a producer has a 50-acre irrigated pasture of orchardgrass, meadow brome, and alsike clover. A reasonable expectation of dry matter yield in the intermountain West is 2.5 tons per acre, or 5000 pounds per acre.

If the producer wants to graze 800-pound yearlings for 90 days, the calculations to figure the stocking rate on an early turn-out to maximize irrigated pasture use is:

$$Number\ of\ animals = \frac{50\ acres\ x\ 5000\ Ib\ /\ ac}{0.036\ x\ 800\ Ibs\ x\ 90\ days}$$

Again, a rapid grazing rotation during the early season is important to consider. At higher elevations, spring temperatures can dip to freezing each night, slowing grass growth. Hitting the pastures too hard too early can impede the system's ability to rebound and deliver good forage production later in the summer. Another approach is to decrease the stocking rate until nights become warmer and forage production begins in earnest. Like any rotational grazing system, controlled grazing in the West requires observation, observation, and more observation. The Chinese proverb holds true here: "The best fertilizer for the land is the footprint of the farmer."

Avoid using irrigated pastures to winter feed hay unless you plan to renovate, drag, or harrow in the spring. Feeding grounds are subject to soil compaction because of the large numbers of animals that congregate there over the winter. Harrowing pastures to distribute manure, although not always cost effective, is often recommended in short-season regions, at least once at the beginning of the growing season. In cold regions with short growing seasons, nutrients cycle in the soil at a much slower rate than in more temperate regions. Manure piles therefore tend to break down slower, and dragging can break them up, increasing surface area and, it is thought, aiding in decomposition.

References

Anon, *The grass farmer's bookshelf,* The Stockman Grass Farmer, June 2001.

Barnhart, Stephen K., *Selecting Forage Species,* University Extension, Iowa State University, Ames, 1999.

Engle, Cindy, *Wild Health: How Animals Keep Themselves Well and What We Can Learn From Them,* Houghton Mifflin Company, New York, NY., 2002.

Gerrish, Jim, *Strategies for pasture improvement,* Forage Systems Update, January 1999.

Gunnink, Doug, *Gross margin analysis helps show the way to grazing profits,* The Stockman Grass Farmer. April 1993.

Leep, Rich, and Doo-Hong Min., *Clovers beat commercial N in Michigan studies,* The Forage Leader, Spring., 2005.

Murphy, Bill, *Greener Pastures On Your Side of the Fence,* Arriba Publishing, Colchester, VT., 1987.

Turner, Newman, *Fertility Pastures and Cover Crops,* 2nd ed. Bargyla and Gylver Rateaver, Pauma Valley, CA., 1974.

5

Dairy Heifer Management

Heifer Selection

Care must be taken to ensure minimum calving difficulty through heifer and sire selection. Do not give heifers creep feed during the suckling period, as fat may be deposited in the developing udder, lowering subsequent milking ability.

The following factors should be kept in mind when selecting heifers.

At Weaning:

1. Select the largest and oldest heifers because they will usually reach puberty earliest. Also, these heifers are usually from the earlier calving and/or heavier milking cows.
2. Select heifers that are structurally large, physically sound, in good health, and have good growth potential. Performance records help to determine growth potential and to identify calves from above-average-producing cows. Do not save heifers that are very fat at weaning because of the tendency for reduced milk-producing ability.
3. If purchasing heifers is part of the replacement programme, the availability of good quality stock is usually greatest at weaning time and replacements can be fed and bred as desired. For producers who wish to maintain a two-breed-cross cow herd and use terminal

sires, buying replacement heifers is a must. With limited feed resources, purchasing bred heifers may be desirable, but their availability will be lower.

As Yearlings:

1. Save more heifers at weaning than actually needed for replacements so the growing heifers can be culled before breeding, and those failing to conceive during a short breeding season can be culled. If heifers can be bred as yearlings and only those conceiving early are kept for replacements, herd fertility can be improved.
2. Save only those heifers that conceive during a 30- to 45-day breeding season. Heifers should be pregnancy checked about 60 days after breeding, and open heifers should be moved to a feed lot for finishing.

A large number of proven and unproven bulls are available for many breeds. In reviewing sires, it is generally accepted that the higher the Predicted Transmitting Ability (PTA), the greater potential to transmit yield productivity to their offspring. Criteria should be developed for purchasing semen to breed heifers. It is best to begin with the proven sires. Their records provide production, type, and calving ease data. The reliability of this information improves as the number of daughters increase. A second mating option is to use the young, unproven sires. Although these bulls have no data for their own progeny, they are the sons of high PTA bulls and their dams had to meet certain genetic requirements. These young sires are selected for testing because of their estimated genetic value. From this group the proven bulls of tomorrow will be chosen. Young sires should be randomly bred throughout the herd.

Although some herds may use 100 percent young sires, geneticists agree that no more than 25 percent of the herd should be mated to young sires. Some caution should be exercised in mating virgin heifers to young sires because the bull's calving ease ability is not known. A benefit of using some unproven sires is the financial incentive. The initial semen cost is relatively low. In conjunction with this, most breeding organizations offer an incentive "cash return" programme for information provided as the calf matures.

With seven out of every ten cows bred artificially, there are still many dairy farmers who consider natural breeding an essential reproduction tool. Natural service generally cannot provide the genetic advancement possible with proven sires. Breeding heifers to an unproven dairy or beef bull retards genetic improvements and reduces the number of herd replacements available. Bulls are dangerous; A.I. eliminates the need to have a potentially dangerous animal on the farm.

Natural service bulls may be subfertile, delaying the age of a heifer's first calving and thus reducing milk production per day of herd life. On the other hand, semen quality and fertility of A.I. bulls are monitored. Suboptimal semen is not distributed. Natural service bulls can also introduce venereal disease into a herd, whereas A.I. sires are disease-free. Conformational type characteristics are generally unknown for natural service bulls. Furthermore, little information is available on size of calves sired by farm bulls or the incidence of difficult births. A.I. sires and calving ease summaries provide considerable information to the dairy farmer about production merit, type classification, and calving ease. With this information readily available, the risks involved in an A.I. programme are much lower than those in natural service.

Additionally, there is a distinct financial advantage to merchandising A.I.-sired heifers. A.I. makes identification easier and promotes the use of more accurate breeding and health records. By using A.I., farmers can better control the time when heifers calve. When planning a mating, it is important to avoid genetic defects. A defect can be present when both of the mated animals are carriers for that particular disorder. When both animals carry the recessive trait, it will appear in approximately 25 percent of the matings. Some of the genetic defects are mulefoot, limber leg, and DUMPS (Deficiency of Uridine Monophosphate Synthase). Bull studs and breed organizations identify bulls which are carriers of these genetic defects. These carrier animals should not be used when selecting the future genetics of your herd. Some of these genetic defects will cause early embryonic death, abortions, or stillborn calves. Others will produce a live calf, however, these animals will never achieve the same profitability as their herdmates.

One of the major decisions with regard to the future calf population is the selection of their parents. This choice is made when calves are either kept as replacements or culled from the herd. Calves that are kept for replacements should be the result of a well thought-out mating with the best bull for a particular cow. The offspring of this mating will hopefully be a heifer that will have greater genetic potential for milk production. To know what kind of offspring will be produced, an Estimated Breeding Value of the heifer can be calculated. This will give the expected mature production of a heifer compared to herdmates in a breed average herd. The example below illustrates the simplicity of this calculation. It is a matter of adding the PTA of the sire and the PTA of the dam.

To obtain the maximum benefit of selection, calf losses must be very minimal. Calving interval, competency of management in raising calves and heifers, and the method used to breed virgin heifers are three of the most important factors determining the number of quality replacement heifers available. Examples of the number of replacements available in a 100-cow herd for two calving intervals, three levels of management, and three methods of breeding heifers are listed in The average herd cull rate is 25 to 30 percent of the cows each year, so a 100-cow herd requires 25 to 30 replacements per year.

A successful calf raising programme will have lower death losses and will provide greater opportunity to cull animals with a lower genetic potential. The lower 10 percent of the calves should be culled to allow for the maximum genetic advancement. However, if calf mortality has been high, all the available calves may be needed for replacements. In this situation, the genetic gain would be limited because inferior animals would not be removed from the herd. Poor reproductive performance in a herd will adversely affect selection opportunities. The potential number of calves available will decline as the calving interval increases.

Accurate records of maternal performance are essential aids in determining which progeny should be raised for replacements. Use of Dairy Herd Improvement records is encouraged, along with records of identification, health, and reproduction. A newborn calf should be properly tagged and identified as soon as possible. The sire and dam should be known and recorded for each calf. The introduction of farm computers and computerized DHIA systems have enabled producers to keep more accurate and up to date records with less time compared to manual records systems.

Heifer Artificial Insemination

Artificial insemination of virgin heifers can contribute the most to genetic progress within a herd. Progeny from heifers (first lactation) account for nearly 32 percent of all births. Despite the disadvantages, potential problems, and safety involved, 40.4 percent of the farms surveyed still had one or more bulls to lessen the inconvenience involved in breeding heifers. Genetic progress is slowed when animals are not bred A.I. Studies have shown that the use of genetically superior A.I. sires results in genetic progress rates three to four times the rate of herds using natural service bulls.

The problems with calf delivery in heifers can be minimized if a common sense approach is taken. Well grown heifers should be mated to proven calving ease sires. Most sire organizations offer a list of calving ease bulls. Some producers use a beef bull either through A.I. or natural service. This also causes calving disorders because the beef breeds are selecting for larger birth weights and larger frame size. It is important to remember that using a beef breed is no substitute for well grown heifers. In addition, a large portion of the calf crop potential is lost. One of the major reasons that more heifers are not bred artificially is inconvenience. This can be resolved with adequate restraint facilities. When re-modelling or building any heifer facility, include restraint and handling facilities for animals of all ages. The genetic potential available from using A.I. on heifers is well worth the time to catch heifers in heat and the investment needed for facilities.

Culling

Herd improvement can be increased if animals are culled for genetic reasons. However, if animals are removed from the herd due to poor management, genetic improvement is severely limited. Average culling rates range from 25 to 30 percent. If cull rates exceed 30 percent, all available heifers will be needed as replacements. Merely having sufficient numbers of heifers available for replacements is not satisfactory. In such situations, genetic advancement is limited because the genetically inferior heifers are not removed from the herd.

Dairy farmers should breed heifers to A.I. bulls with top production proofs and good calving ease ratings. This allows them to select which replacements will enter their herd and which ones will be sold to other farmers (needing additional heifers to maintain herd size). Also, with plenty of genetically superior replacements available, some inferior cows could be culled that otherwise would have been tolerated. In determining the number of herd replacements needed, it is necessary to consider the effect of age at first calving. For each month after 2 years, the herd requires an increase of 4.2 percent in the annual replacement rate. By calving at the recommended age of 24 months, fewer replacements are needed.

Calving Time

Precalving Heifer-dry Cow

The care and feeding of heifers or dry cows does affect their unborn calves. Size of the fetus and calving ease are determined by genetic factors, nutrition, age, and the size and condition of the cow or heifer at calving time. Dry cows should be fed separately from the rest of the herd and should be in good body condition. Calving problems can result when cows are thin or fat, heifers are undersized or oversized, or calving facilities are poor. The results include stillbirths, injured cows or calves, and postpartum reproductive complications.

Cows should be moved to a closed maternity pen or box stall a week or several days before their expected calving date. Cows and their newborns need protection from the other cows. Clean, dry maternity pens should be at least 12' x 12' in size. The stalls or pens should be well lighted and ventilated but free from drafts. It is best to clean the pens after each use and apply lime or other granular material to the floor before covering it with adequate amounts of dry bedding such as clean straw.

A good nonslip base and adequate amounts of bedding can prevent injuries and udder trauma during calving. Wet sawdust, moldy hay, moldy, damp silage, or spoiled haylage should not be used for bedding. Many cases of infectious mastitis can be traced to contaminated bedding, especially green wood shavings and sawdust. Moldy hay and silage, and manure contaminated bedding contain organisms that can infect the uterus and udder. These organisms can also infect calves. During mild, dry weather, a well drained paddock or small pasture with shade can also serve as a good calving area.

Care at Calving Time

Approximately 2 to 5 percent of all calves are born dead (stillborn), many of which could have been saved if someone were present at birth to render proper assistance. Too often cows get assistance only after the cow or calf is in critical condition. The value of cows and calves makes it very profitable to observe the cows frequently prior to calving. Cows should also be attended to during calving. Monitoring the birth process ensures that cows and their unborn calves receive assistance if it is required.

North Carolina University researchers estimate that each difficult dairy birth costs \$40 to \$75 due to increased calf and cow mortality, reduced milk production, re-breeding problems, and the cost of additional labour required in delivery. Proper management of dry cows and heifers, and using sires with good calving ease ratings can help reduce these losses. Cows should be checked every 2 hours after the onset of labour to monitor progress.

Some cows show noticeable signs of calving and will give birth in a few hours, while others may not calve until the next day or may calve earlier than expected. Signs of discomfort usually appear when the cervix has begun to dilate. Arching of the back is apparent at this time. Definite straining does not occur until the chorioallantois (outer placental sac) approaches the vulva. Pressure from the fetal fluids moisten this membrane and help to complete the dilation process. Contractions become more intense as the fluid sac ruptures. Then there is a temporary weakening of abdominal contractions until the ammion (inner fluid sac) enters the vulva. Once this membrane ruptures, regular contractions and straining begin to increase in frequency and duration until parturition.

Cows should deliver 30 minutes to 1 hour after the fetus appears in the pelvic inlet, and heifers should deliver within 2 hours. If the cow or heifer does not progress accordingly, she should be examined to check for abnormal fetal position or other problems. Cervical dilation must occur before assistance is provided.

After Calving

The cow should get up and assist the calf within 30 minutes after giving birth. If she does not getup soon after birth seek advice and assistance from a veterinarian. The cow should pass manure and drink water within this time. A normal cow will be alert, have a normal body temperature, and be willing to eat and drink within an hour or two after calving. The cow should be allowed to lick the calf after delivery. Licking stimulates the calf's blood circulation and may increase absorption of immunoglobulins in colostrum.

Management after Calving

Each calf should be positively identified before it is removed from its dam. This is required if calves are to be registered and it is essential for good breeding programme records. A neck strap or chain with a number or a metal or plastic ear tag can be used for identification. The ear tag or registration numbers of calves, sires, dams, and the birth dates should be entered in a permanent record book.

Permanent identification methods include photographs, sketches, tattoos, and freeze branding. The Holstein, Guernsey, and Ayrshire breeds require a photograph or sketch for registration. The other breeds, Jersey, Brown Swiss, and Milking Shorthorn, require tattoo markings inside the ear. The Ayrshire breed accepts both types of identification. The tattoo or freeze brand also provides permanent identification of dairy heifers of any breed for farm use.

Birth through Weaning

Feeding

Feeding young dairy calves is critical to raising replacements. During the first 2 months of life, dairy calves function primarily as a monogastric (simple-stomached animal). After about 2 months of age, they begin to function more like a full-fledged ruminant. During these first few weeks of life the rumen, reticulum, and omasum of the calf are relatively small in size and are quite inactive compared to the abomasum or "true stomach." For this reason, young dairy calves have special requirements for protein, energy, and vitamins.

Newborn calves cannot utilize vegetable protein before their rumen is functional because they have limited digestive enzymes. Therefore, following colostrum feeding, whole milk, fermented colostrum, or milk replacers containing milk protein or specially processed soy concentrates should be used. By the time calves are weaned, they can utilize most vegetable proteins very efficiently.

Young calves cannot digest starch or some sugars such as sucrose (table sugar), because certain digestive enzymes are not present. Calves are limited by the type of fat they can utilize but can digest saturated fats such as milk fat, coconut fat, lard, and tallow. They are less able to digest unsaturated fats such as corn oil and soybean oil.

Major sources of energy for newborn calves should be derived primarily from lactose (milk sugars) and milk fat. It is important that calves have adequate energy because the metabolic rate (rate at which energy is used) is greatest during the first two weeks of life. Within two weeks, the calves develop the ability to digest starch. Shortly thereafter, they develop the ability to digest complex carbohydrates. The rate of rumen development dictates how rapidly young calves can digest complex starches and carbohydrates.

Vitamins required by calves include the water soluble B vitamins (thiamine, riboflavin, niacin, choline, biotin, pyriodoxine, folic acid, B12, and pantothenic acid) found in colostrum, fermented colostrum, whole milk, or good milk replacers. Rumen microorganisms are able to produce these when the calves' rumen begins to function. Calves require the fat soluble vitamins A, D, and E; they are in short supply at birth but are found in colostrum. Whole milk, fermented colostrum, or milk replacers plus some sunshine will normally supply an adequate amount of these vitamins to young growing calves.

Dairy calves require the same minerals for growth as do other animals. Milk, fermented colostrum, and milk replacers generally supply adequate amounts of minerals necessary during the first few weeks of life. A significant point to remember is the mineral content of colostrum and milk may be low or deficient depending on the mineral status of the lactating animal's diet. Selenium may be reduced in milk because of a dietary deficiency, therefore, mineral supplements are needed for young calves.

Water makes up the majority of a young animal's bodyweight. It is the media in which all chemical reactions in the body take place. A plentiful supply of water is needed for normal rumen fermentation and metabolism, digestion, and absorption of nutrients, as well as excretion of waste products. Water quality and it's availability to all animals are also equally important. Limiting good quality water to dairy calves and heifers may limit growth and also adversely affect the health of these animals.

Several factors which can influence intake of water are dry matter intake, water content of the diet, environmental temperature, and mineral-salt intake. To better understand the nutrient needs of young calves it is necessary to know their nutrient requirements. The requirements for large and small breed calves are presented and within each section are the animal's weight, expected gain, dry matter intake, energy for growth, and protein requirements.

Health

Several management, environmental, and physiologic factors can affect young calves making their life more difficult. The first place where good management can contribute to good health care is at calving. A Pennsylvania survey indicated management factors do affect calf mortality. The practice of washing udders after calving and before calves nurse or colostrum is milked, as well as the practice of using frozen colostrum and assisting the calf with its first feeding, influence calf mortality. It is apparent that proper management, feeding, and sanitation help ensure that baby calves have enough immunoglobulins to combat disease causing organisms.

Calf scours during the first month of life are the most common cause of calf sickness and death. Several pathogens, including E. coli, rotavirus, coronavirus, salmonella, coccidia, and cryptosporidia, cause calf diarrhea. These agents can be present year-round, but under conditions of crowding, cold stress, inadequate nutrition, and less frequent manure removal, calf scours can become a herd problem. The best cure for scours is prevention. Calves should be fed correctly and housed in a clean environment. Calves with scours should be treated immediately with a homemade or commercial electrolyte solution to keep them from dehydrating. Milk should not be fed when scours occur because milk may encourage growth of bacteria in the intestine and further complicate the scours. Milk or milk replacer should be replaced with an electrolyte solution for no more than 24 to 48 hours.

Feed daily at the rate of 1 pint (approximately 1 pound) per 10 pounds of bodyweight to calves afflicted with scours. Used with the usual scour treatment, the solution will prevent or alleviate dehydration. Feed three to four times daily during a 1-day or 2-day perlod as a substitute for the usual milk or liquid milk replacer. An internal parasitic infection becoming more of a problem in recent years is coccidiosis. Coccidia are single cell protozoa that live within the cells of the digestive tract. After a coccidia infection has begun in the animal, the coccidia organisms can also spread to various locations within the intestines.

These organisms have the potential to cause extensive damage to the intestinal tract of young animals. Calves often become infected when between 3 and 6 weeks of age and while confined in pens, although older calves weaned in confined lots also show symptoms of coccidiosis. Infected calves show signs of bloody diarrhea and may become dehydrated and die. Coccidiosis at the subclinical level (undetectable by usual clinical observations), reduces the growth rate of calves. With time and treatment, animals develop an immunity that keeps the numbers of these organisms at low levels.

As calves get older they are still subject to diseases. At 6 to 8 weeks of age, respiratory disorders seem to be the largest problem and are often associated with high population density and inadequate ventilation. In closed-up calf and maternity barns, germs, toxic gases, and moisture all accumulate. Respiratory disorders are accentuated with high relative humidity at low environmental temperatures. Respiratory diseases, such as pneumonia, tend to be worse during winter and early spring.

Research has shown a high correlation between respiratory disease and calfhood morbidity, growth rate, as well as reproductive efficiency, and average age at first calving. The immediate causes of pneumonia are bacteria and viruses, but of greater significance are the predisposing factors of dampness, drafts, chilling, and toxic gases. The accumulation of gases in confinement housing irritate the respiratory tract. Ammonia is one of the major irritants. When the smell of ammonia is noticeable, a high probability exists for damage to the lung defense mechanisms of these confined animals.

Calf mortality also tends to increase during cold, wet, and windy winter weather. This occurs because maintenance energy requirements are much higher in a cold environment; antibody absorption from colostrum is less efficient during winter; and lice, mange mites, and ringworm fungi prosper in a confined damp environment and under conditions of poor nutrition. Therefore, observation of young calves on a regular basis is critical so their diets may be altered, if necessary, to sustain energy levels during these types of conditions.

Health management for dairy calves during cold weather means more intense management procedures. Some tips to optimize the care of young calves during winter conditions are to pay closer attention to dry cow care, calving-time care, colostrum intake, feed quality such as a highfat milk replacer (greater than 10 percent fat), environment, and checking calves at least twice each day. Keeping on top of health care management during this stressful period will increase the chances of producing healthy calves.

Two important health management practices that can be implemented during the preweaning period that will help alleviate some stress at weaning are dehorning and extra teat removal. Since horns and extra teats are of no value to dairy animals, they must be removed to help prevent injury, to improve their opportunities for continuing health, and to improve their appearance. The ideal time to dehorn calves is between 2 and 3 weeks of age or as soon as the horn button can be felt. Calves need to be restrained to prevent the calves or the operator from being hurt.

Heavy-duty electric dehorners are one of the most humane, effective, and safest tools to use. When electrical dehorners are used correctly, a continuous copper-coloured ring will be displayed around the base of the horns. The surface of the iron should be cherry red before it is touched to the horns. This results in a very minimum of pain to the calves and provides very rapid destruction of the horn buttons. A second application of the iron may be necessary if a spot has been missed. When calves are restrained for dehorning, they are easily examined for extra teats. Examination and removal are easier when the calves are lying on the floor. The four regular teats should be symmetrically arranged with the two rear teats set slightly closer together. Extra teats are usually smaller in size and located close to the main teats. They can be snipped off with a pair of scissors. The cut should be made lengthwise with the body and painted with iodine or another topical antiseptic. If you are not absolutely sure which teats are the extras, do not remove any of them. Allow your veterinarian to make the decision.

Housing

Housing often is the weakest link in a calf and heifer operation. While a good calf raising facility must minimize environmental stress on the calves, it must also minimize environmental stress on the part of the operator. The latter encourages the operator to maintain a higher level of calf care. The calf raising facility should be comfortable for the calves and convenient for the operator.

General housing requirements for pre-weaned calves are individual housing for each calf, isolation from older animals, well-ventilated but draft free quarters, dry pens with ample bedding, and a suitable location to encourage regular observation. These factors will help ensure that calves will start off strong and healthy. A proven housing system for dairy calves in all types of weather is the portable outside hutch located in a well-drained and protected area with the open-front facing south.

A 4' x 8' individual hutch or similar sized solar-type calf hutch is recommended. Only one calf should be placed in each hutch. Use the equivalent of at least a bale of straw in each hutch. Clean and disinfect hutches or let them stand upright in the sun for a period of time after each calf has been removed. Move the hutch to a new location and place fresh fill and bedding under the hutch prior to its reuse. An alternative to outside portable hutches is to use a new or existing cold, but well ventilated, barn with pens at floor level. Pens should be 4' by 8' with three solid sides and a slatted opening on the feeding end. Pens should be removable to allow for their cleaning, disinfection, and the cleaning of the floor between pens. If smaller pens are used, it is necessary to clean the pen during the grow-out period (month 0 to 2) to prevent buildup of manure and associated gases and odours. Portable pens can be used in more than one building or area of a building as a way to break disease cycles.

Another popular housing method is to incorporate calf-hutch type structures in a building with three sides enclosed. This method can be modified to allow calves to have outside runs. This type of housing allows the operator to be under cover when feeding and working with the calves. Forty-four percent of farmers use calf hutches for all or part of the year. The number of farmers using calf hutches is almost equal to those keeping calves in dairy barns with older animals.

Effectiveness of ventilation in baby calf and weaned calf housing influences calf mortality. A good ventilation system should allow adequate air movement, eliminate the accumulation of gases such as ammonia, and keep humidity levels at a minimum. For these reasons, natural ventilation is recommended for baby calves. Isolated, heated, mechanically ventilated calf barns are not recommended.

The number of farms that reported mechanical ventilation systems was greater than those that reported having a fan. Fewer farmers reported having the basic components of a natural ventilation system than reported having a natural ventilation system. Many farms had no planned or functioning ventilation system but still referred to it as a natural ventilation system.

Weaning Dairy Calves

Weaning time is a critical and stressful period in the life of dairy calves. Management practices and changes made at the time of weaning can have major impacts on their immediate and possibly long-term future. Situations do exist where weaning day consists of a complete change from some milk, grain, and hay to a ration of all grain, hay, and possibly silage; a move in location; progressing from being pampered to fending for themselves, and even fighting to get to the feed bunk. Add to this being de-horned, vaccinated, having extra teats removed, and getting one or more new ear tags; it is easy to see that problems could develop. Calves often lose weight for several days following weaning.

Management situations often make it necessary to implement many of these changes within a short period of time. However, the following practices can help reduce the stress placed on these calves before, during, and after this weaning period. A small-group housing facility for three to five calves should be used for at least a month immediately after the post-weaning period.

This post-weaning area should be similar in housing type and environment compared to the grow-out facility, and located in a similar area. This allows for regular feeding and observation of the 2-to-3- month-old group while feeding the younger calves. Large super hutches located adjacent to a calf hutch area, or an open front shed with an outside yard adjacent to the calf hutch area, can be used in conjunction with small hutches. Buildings with inside individual pens should also provide for a group pen at least 12' x 12' for three to five calves. Pens should have a continuous supply of frost-free water. At least 18 inches of bunk space should be provided for each calf, with stanchions or dividers to define the eating positions.

Delay moving calves until 2 to 4 weeks after weaning. Calves will be under less stress at this time and suckling problems are less likely to occur. Calves should not be moved from a warm to a cold environment or from a cold to a warm environment at weaning. Other management procedures to avoid stressful conditions at weaning are making sure that calves are drinking 8 to 10 quarts of water each day. Feed the same type of forage and grain before weaning as calves will get after weaning, and make sure they are eating adequate amounts of these before weaning.

Calves should be eating at least 1.5 pounds of calf starter per day before they are weaned. Those who are slow in adapting to grain could be kept on liquid feed for a longer time. Also make sure that calves are eating some forage so their systems are adapted to dry forages. Because forage should be offered at 7 to 10 days of age, most calves will be eating enough by weaning time. Water is important to increase intakes of dry grain and forage, and these are important to increase the development of the rumen.

An optimum situation is weaning three to five calves of similar ages, moving them to a location of similar environment with adequate feed bunk space and easily accessible water, maintaining the same forages and grains as were fed before weaning, and invoking little if any other change in management practices during that time. The best situation does not always exist, however, the closer it is achieved, the less stress the calves will undergo. Observation of recently moved and weaned calves should be done frequently during the first few days after weaning. It is essential each calf be eating feed and drinking water. Some calves that are shy and easily pushed out of the way may need some extra care to get them through this critical period.

Weaning through Six Months of Age

Feeding

When weaning time arrives and the liquid feed portion of the young calves' diet is reduced, it is necessary to provide them with adequate substitutes. Those substitutes should be a good quality free choice grain mixture and an ample supply of high quality forage. A fresh, abundant, and clean supply of water also must be available before, during, and after weaning because inadequate water will reduce the animals' grain and forage consumption. It is essential that calves and older heifers are maintained on a high plane of nutrition so that growth will continue at a normal rate.

Calves are able to function as full-fledged ruminants after 4 and 6 months of age. Developing a fully functional rumen is an important part of early calf nutrition. A developed rumen allows calves to fully digest and utilize forages and dry grains, and to be fed a lower cost ration than was previously possible. This creates the need to begin grain and forage feeding very early. Heifer nutrition is all too often neglected between weaning and 6 months of age. During this time period, calves are in a transition from high quality feeds used before weaning, to forages and grains that are of a lesser quality, especially in terms of protein.

High-quality forages (alfalfa hay or other legume/grass hays) and sufficient amounts of a grain mixture need to be offered to young calves. A maximum of 4 to 6 pounds of grain per head per day should be offered to young heifers up to age 6 months. This grain mixture should contain 16 percent crude protein in most situations. Immediately after weaning, young heifers consume a very small amount of forage dry matter in comparison to the amount of grain mixture consumed. Carefully monitor forage quality during the period when heifer calves are age 2 to 6 months. Fine stemmed, mold-free hay should be fed.

A good quality second or third cutting legume grass mixture hay is preferred. Poor quality, stemmy, or moldy hay reduces forage intake and holds back growth. As animals get older and larger, their forage intake increases. By age 4 to 6 months, it is important that heifers are eating palatable and high quality forage. Pasture is not recommended for very young calves. However, it can be used for part of the feed for calves that are between 4 and 6 months of age. Careful monitoring of the pasture and ample grain supplementation is necessary. It is beneficial to keep a supply of free-choice hay on hand in an area where calves may seek shelter on hot days or when flies become a problem. Regardless of the type of forage fed, young heifers need a supplemental feeding of grain. The amount of grain fed will depend upon the age of the animals and on the quality of forage being fed.

Total mixed rations can also be fed to heifers after 2 months of age. Feeding complete rations encourages heifers to consume several small meals during the day and leads to better feed digestion and utilization. These feed rations need to be carefully balanced and they normally need to have mainly dry feed components. The advantages of total mixed rations for these heifers are that all the needed ingredients are in the desired proportions, time and labour are reduced, and competition among animals is minimized. In addition to the quality and amounts of forage and grain being fed at this time, it is also essential to know these calves' nutrient requirements. Within each weight classification for these growing animals, the requirements are listed for three rates of gain to allow for variable growth rates under different environmental and economic situations.

Health

Many vaccines are available for Pennsylvania dairy farm replacement animals. Some of the most commonly used vaccines are for brucellosis, IBR (infectious bovine rhinotracheitis), PI3 (parainflu enza-3), Roto-corona, and colibacillosis. Some other available vaccines protect against pasturella, pinkeye, haemophilus, leptospirosis, vibriosis, and clostridia. The specific needs of each farm should be discussed with the herd veterinarian. All heifer calves should be vaccinated for brucellosis when they are 4 to 8 months of age. It is also a good idea to vaccinate all replacement heifers against IBR, PI3, BVD (bovine virus diarrhea), and BRSV (bovine respiratory syncitial virus). Six to 8 months of age is a good time to do this because maternal immunity has diminished by this age. In herds with widespread respiratory infection in young calves, a nasal vaccine to protect against IBR-P13 may be given to calves as early as 2 weeks of age.

Internal and external parasites can be serious problems affecting the growth and performance of dairy heifers at any age. The control of internal parasites in young calves requires careful attention to clean housing and management practices. Young calves should be housed individually to eliminate contact with infested animals or manure until at least a week after weaning. Calf facilities should be kept free of manure buildup. Most herds require a routine treatment programme for young calves from weaning to 8 months of age. Serious internal parasite problems are most likely to occur during the animals' first season on pasture.

Heifers on pasture can be heavily infested with parasites by early to mid-summer. Parasitized heifers grow more slowly, have lower feed efficiency, may develop diarrhea, and become anemic. Strategic worming at 3 and 6 weeks after turnout on pasture, and in the fall after confinement again, will greatly reduce the degree of internal parasitism. External parasites of concern to dairy farmers in Pennsylvania are lice, mange mites, and flies. These can all reduce growth and performance of calves and heifers. The spread of all parasites should be limited by preventing direct contact between age groups of heifers and by cleaning all pens thoroughly before putting a new group of animals in a pen. Sanitation and good management are just as important as drugs in a calf and heifer parasite control programme.

Housing

Once weaned calves have adjusted to their group pen and are growing, they can be moved into the first stage of a multi-stage calf grow-out facility such as a gated free stall facility, a counter slope facility, a gated bedded pack facility, or a combination of these. This move might be made when the calves are as young as 4 months or as old as 6 months. They may be placed in groups of 10 to 12 with age and size spreads no more than 2 months or 100 pounds within each group. Allow about 25 square feet per calf, 18 inches of feed space, and a continuous supply of water. If calves are located in a multi-age facility, their contact with older animals should be minimized by use of a solid partition. These requirements are best met by a series of free stall areas, bedded packs, or pens along a feed bunk.

Proper ventilation, especially control of humidity, ammonia, and other odours is recommended for animal comfort and health. High relative humidities have been associated with poor performance and high incidences of respiratory diseases in confined calves.

Six Months of Age through Calving

Feeding

Most forages, with the exception of silage, can be offered to heifers on a free-choice basis without causing an overeating problem. For this reason many dairy farmers keep forages available for their heifers at all times. Hay is most often available, while silage must be replenished at least once per day to avoid spoilage and reduced intake. The inconvenience associated with daily silage feeding can be reduced with a bunker silo equipped with a moveable electric fence or self feeder. With corn silage, anhydrous ammonia can be added at the time of silo filling to reduce the mold growth and increase the bunk life of this feed. Anhydrous ammonia treated corn silage can be fed on an every-other-day basis with less concern about the silage heating or spoiling.

Pasture also makes an excellent forage for heifers when an adequate amount is available. In the hot dry parts of summer, supplemental forages must be fed to provide adequate amounts of dry matter. Forage intake for heifers should be 2.0 to 2.2 pounds of dry matter per 100 pounds of bodyweight.

Abundant pasture that is kept young and actively growing may provide most of the nutrients needed by heifers more than 9 months old. Pastured heifers should be provided with a minimum of free-choice dicalcium phosphate and trace mineral salt. If grass tetany or sudden deaths are a problem on pasture, supplemental magnesium should be provided in a manner to ensure intake of at least 1 ounce per head per day. Supplemental forage should be provided only when necessary to help control bloat on legumes or to make up for a lack of pasture. Limited grain feeding is necessary to provide a vehicle for bloat or tetany-control additives, to increase energy intake when pasture is scant or overmature, and to serve as a way to incorporate needed minerals, vitamins, and other feed additives, such as ionophores (monensin or lasalocid), into the diet.

Priorities for feeding young stock during this time are to supply a balanced diet on which they can grow and stay healthy so they can be bred at 13 and 15 months of age and maintain their pregnancy. To ensure heifers are receiving their required nutrients, it is advisable to test the forages that are being fed and to balance a ration for animals 6 through 12 months and those over 12 months of age. The effort put into feeding a balanced diet will pay off when these heifers eventually freshen and enter the milking herd.

Growing heifers often requires some grain. The amount of grain fed and the concentration of nutrients in it are determined by the average forage consumption and the average nutrient requirements for the group of heifers. This grain mixture is usually offered daily and the entire amount is consumed within a short period of time. Heifers being fed good-to-excellent quality forage need 1 to 3 pounds of concentrate per day, while those on poorer quality forages require 3 to 5 pounds of concentrate each day. The percentage of protein and other nutrients vary according to the forage being fed.

Under many management conditions, dominant heifers consume more than their proportionate share of concentrate, resulting in an excessive variation of individual growth rates within groups. These problems can be reduced by sorting animals into more uniform groups by size and/or having some form of locking head gate to allow all animals the opportunity to receive their proper amounts of concentrates. All heifers should be able to be at the feed bunk at the same time.

Total mixed rations are ideal for heifers during this age group. When fed this way, heifers are allowed to consume rations free-choice with the fiber and/or bulk of the ration used to regulate intake. Since rations are available at all times, all animals are able to obtain adequate dry matter.

Health

The major health problems of young stock are internal parasites, external parasites, and infections causing respiratory problems and abortions. Any of these disease problems may cause significant damage before detection. It is definitely more cost-effective to prevent the problems than trying to contain the damage.

Most dairy farms have a routine treatment programme for young stock. Worming should be started soon after calves are put into group pens. Manure samples from each group or pen can be examined by the herd veterinarian for presence of worm eggs or coccidia oocysts. This helps determine the need for worming and monitors the success of the worming programme. External parasites should be treated in fall or early winter. Several drugs are available for treatment of stomach and intestinal worms. Lungworms, when present, can be treated with levamisole. Coccidia are not affected by wormers, but can be controlled by one of several available coccidiostats.

External parasites, like the internal parasites, will cause serious production losses in young stock. The major external parasites of concern in Pennsylvania dairy herds are lice, mange mites, stable flies, house flies, face flies, horn flies, and heel flies. Any of these can be serious enough to affect growth rate and feed efficiency. As mentioned earlier, pastured heifers should be wormed and confined heifers may also need regular worming due to parasites that may be picked up in the dry lot, manger, and from other animals. Manure samples should be checked by a veterinarian for recommendations on a worming schedule.

Vaccines against diseases such as pinkeye and haemophilus may need to be considered in some herds. Heifers should be vaccinated against leptospirosis at least 30 days prior to breeding age. A vibrio vaccine should be considered if natural service is used. An important element of a heifer health programme is to have restraint facilities (chutes, corrals, self-locking stanchions) for all heifers, especially those of breeding age.

Heifer Reproductive Management

Heifers have reached puberty when normal sexual behaviour is exhibited and ovulation occurs. Onset of puberty more closely relates to bodyweight than to age. Heifers reach puberty when bodyweight is 30 to 40 percent of the average adult weight, and should be ready to breed around 13 to 15 months of age.

Puberty is delayed if growth is slowed by underfeeding, disease, or parasites. Low dietary energy levels can lead to ovarian inactivity. Inadequate protein intake and nutritional problems leading to anemia can cause silent or irregular heats. Deficiencies of phosphorus, vitamin A, and vitamin E may also affect reproduction. Heifers approaching breeding age should be watched closely for heat to ensure that they are cycling. A veterinarian can examine heifers to determine those that are cycling and also identify those with congenital reproductive abnormalities.

A successful A.I. programme involves routine heat detection and timely insemination. Considerable variation exists between animals. The average interval between heats for heifers is 20 days. All heat dates should be recorded on a chart so that future heats can be anticipated. To monitor heats accurately, dairy farmers must clearly identify heifers with neck chains, large ear tags, or freeze brands, and check animals for heat behaviour twice a day. If heat detection is routine and frequent, heifers should be inseminated 12 hours after the beginning of standing heat. When the onset of heat cannot be accurately determined because of infrequent heat detection, heifers should be bred soon after standing heat is noticed. Keeping accurate records of the breedings is important for predicting calving dates.

Dairy farmers who cannot routinely check for heats may consider using heat synchronization and a concentrated A.I.

breeding programme for selected months during the year. Heat detection can be made easier and more efficient by use of heat mount detectors, crayon or chalk markings, surgically altered bulls, and/or androgenized heifers equipped with chin-ball markers.

Facilities should be provided where heifers can be confined for close observation or until they are individually restrained for breeding or examination. For a relatively small investment in time and money, most heifer facilities can be upgraded and equipped to provide convenient restraint and efficient handling of heifers. Such facilities can also be used for pre-breeding examinations, vaccinations, worming, pregnancy examinations, estrous synchronization, and possibly embryo transfers.

Precalving

Bred heifers can be fed and handled in the same manner as other yearling heifers until about the last 3 months of pregnancy, when the unborn calf makes nearly two-thirds of its growth. During the final 3 months, bred heifers may need extra nutrients to maintain proper body condition for their first lactation and support for their growth plus that of the fetus. Heifers being fed excellent-to-good quality forages (forages containing 60 percent or more TDN on a dry matter basis) should receive 2 to 4 pounds of concentrate per head per day. This should be balanced according to the protein and mineral needs of the animal. Those heifers being fed fair-to-poor quality forages (those containing less than 60 percent TDN on a dry matter basis) should be given 4 to 6 pounds of concentrate per head per day.

Improper feeding and management practices often result in undersized heifers. The common cause is underfeeding forages as well as feeding an unbalanced grain mixture, especially in the areas of protein and minerals. The problem often is solved by access to better pasture and to supplemental feeds. Slowing growth below recommended levels is unprofitable because it eventually shortens the production portion of the heifers' lives. The result of underfeeding heifers is reduced growth and delayed first calving (greater than 26 months). Stunted growth will also result in smaller and less productive cows. More calving difficulties are encountered with undersized heifers than with those that are well grown.

Accelerating the growth of heifers to the extent that they become fat is also undesirable because lifetime milk production and longevity decrease. Studies have shown that excessive intakes of energy (140 percent of recommended levels) before breeding result in fatty infiltration of the mammary gland and reduction of the number of alveolar cells available for milk synthesis. Overconditioned or fat heifers often are the result of overfeeding high quality forages, especially corn silage, and in some cases is caused by excessive feeding of concentrates.

About 30 days before calving, bred heifers should be moved to a clean and dry environment. Lack of this clean environment can cause heifers to have mastitis and high somatic cell counts. If possible, it is a good idea to house these heifers with the milking herd. Permitting heifers to become accustomed to the new surroundings of the milking herd as well as the milking parlor, if that is the case, will enable them to deal with some of the new stresses they will have to face in their early weeks of lactation.

It is important to avoid high intakes of either corn silage or legume forage during this time. Grain intake should be gradually increased to reach a level of about 0.5 percent of bodyweight daily. If lactating cows receive nonprotein nitrogen (NPN) in their diets, heifers should also receive it throughout the prefreshening period. This practice ensures that their digestive system will be well adapted to this source of protein. It may be important to limit minerals, especially salt, if caked or congested udders are a major herd problem. Balancing and evaluating the overall ration may be necessary in such cases.

Much can be learned about the success of a calf and heifer rearing programme by measuring height and weight of these animals. Most dairy farmers, extension agents, feed industry people, and veterinarians are able to recognize underconditioned or overconditioned animals. However, few can judge by sight whether a heifer's height or weight is normal for her age. Measuring and weighing allow a comparison to standards or breed averages and can indicate some problem areas that should receive attention.

In general, the variability of weight increases with age. The overall goal of a heifer raising programme should allow for a relatively constant rate of growth. Holstein heifers should reach 750 to 800 pounds and 48 to 50 inches by the desired breeding age of 13 to 15 months. Beyond this, heifers should be calving at 1137 to 1296 pounds and measuring 52 to 54 inches tall when they are 24 months of age. While every heifer may not conform to these standards, the majority of heifers should be somewhat near these standards in order to be large enough to breed at 13 to 15 months of age and, subsequently, to calve at 24 months of age. The only real way to tell how heifers are growing is to weigh and measure them several times a year. Once or twice a year is better than not at all.

The materials needed to weigh and measure calves and older heifers are a weight tape, a measuring stick, a piece of paper, and a pencil. It works best with three people: two to do the weighing and measuring and one to do the recording of numbers. Twenty animals per hour can be done in any reasonable restraining facility.

Some important points to remember when taping animals:

— Make measurements with the animal standing straight on a level hard floor surface and with weight equally balanced on all feet.

— Watch for excess manure and dirt on the underside of the heifers which could bias the tape measurements.

Costs of Raising Dairy Heifers

Raising heifers for replacements is an expensive item on dairy farms. Often many homegrown items are not considered in the actual costs involved in raising heifers, although they should be. Few farmers realize the true costs of raising heifers.

Reducing Replacement Costs

The largest expense incurred when raising calves and heifers is feed costs. Feeding for optimum growth accounts for about 53 percent of the rearing cost. Starting with the baby calf, some ways to lower these costs include feeding discarded or surplus colostrum and milk replacers instead of saleable milk. Colostrum is often wasted and it can be fed to calves. Overfeeding of liquids can also be costly as this tends to keep the calves full and inhibits their desire for forage and grain.

Grain mixes containing by-product ingredients can be fed to calves and heifers without sacrificing quality. Examples of by-product ingredients include bran, corn gluten feed, or dried brewers grain. Ear corn tends to be a better buy than shell corn, and using some oats or barley can lower the cost of a grain mix.

Forages fed to all age groups should be of the appropriate quality and cost. Fine, stemmy hay should be fed to pre-weaned calves with good quality forages being fed to calves until 12 months of age. Poorer quality forages can be utilized beyond 12 months; but only if proper grain rates are maintained. In addition, feed refusals from lactating cows can be incorporated into a heifer ration.

Forage testing and feed programmes need to be examined for all heifer groups, ranging from 0 to 6 months, 7 to 11 months, and 12 to 24 months. This helps ensure that heifers are not going to become overconditioned or underconditioned.

References

Fiez E., "Sizing up heifer age", *Dairy*, January, 1989.

Heinrichs A.J, et al, "Standards of weight and height for Holstein heifers", *Journal of Dairy Science*, 1987.

James R.E, et al, "Heifer feeding and management systems", In H. Van Horn and C. Wilcox, (eds): *Large Dairy Herd Management*, American Dairy Science Association, Champaign, Ill. 1992.

6

Milking Systems

Milk and dairy products were in short supply and for the most part unavailable to those not living on or near the farm. Milk production was seasonal, creating periods of excess as well as deficiency in the family milk supply. Stabilisation of these production fluctuations by storage and/or further processing into butter, cheese, or other milk products was precluded by the lack of refrigeration.

Consequently, marketing of milk, butter, and cheese was limited to towns which could be reached by horse-drawn wagons. Over the years modern technology has rectified these problems and today a wide array of safe, wholesome dairy products are available to people of most of the world.

Production of Milk

Production of quality milk is the concern of:

— consumers of dairy products
— retail distributors (super markets)
— milk and milk product processors
— dairy cooperatives
— state regulatory departments
— veterinarians, and
— dairymen.

From the list it's obvious that very few of us are left out. Whether we derive a living from the dairy industry through employment or otherwise, most of us are at the very least consumers of dairy products. In the sections to follow we will examine some of these consumer interests and the efforts made in dairy product processing (from the farm to the retail shelf) to preserve the public's confidence and safety.

Composition and Nutritional Value of Milk

Milk is the lacteal secretion, practically free from colostrum, obtained by the complete milking of one or more healthy cows. Nearly 12% of the American household's total food expenditure is for dairy products. Milk and milk products alone provide 10% of the total available calories in the United States food supply, and in addition, represent one of the best natural sources of essential amino acids for human nutrition. These nutritional attributes of milk have long made it a mainstay particularly in the diet of growing children. There are estimated to be some 8 to 10,000 different types of milk products available thus making it an exceptionally versatile raw product.

Milk is composed of water, fat, protein, lactose and minerals (ash). The concentration of these components will vary between cows and breeds. Total milk solids refers specifically to fat, protein, lactose and minerals. This is to be differentiated from solids-not-fat milk (SNF), a frequently used term which describes the total solids content minus fat. SNF milk is known to most people as "skim milk". The nutritional as well as economic value of milk is directly associated with its solids content. The higher the solids content the better its nutritional value and the greater the milk product yields. For example, cheese yields are directly related to milk casein content.

Flavour and Odour

Consumer acceptance is greatly affected by flavour. There are several factors which may produce off-flavours and/or odours in milk. Some of the more common causes of flavor and odour problems are:

- Feed and weed flavours
  - wild onion or garlic
  - strong flavoured feedstuffs such as alfalfa silage
- Cow-barny flavours - which result when milk is obtained from unclean or poorly ventilated environments, improperly cleaned or sanitised milking equipment
- Rancid flavours - presence of free fatty acids (FFA)
  - due to excessive agitation of milk during collection or transport
  - breakdown of the milk fat component by proteolytic and lipolytic enzymes present in raw milk
- Malty flavours, high acid flavours
  - bacterial contamination
- Oxidised flavours
  - exposure of milk to sunlight
  - contact of milk with oxidising agents such as rust, copper, and chlorine
- Foreign flavours
  - fly sprays, medications, etc.

Milk Processing

A multitude of events take place in the process of delivering milk from the farm to the dinner table and all are designed to provide the consumer with a wholesome, nutritious and safe product. The production of quality milk and milk products begins on the farm and continues through further handling, processing and distribution.

Milk processing has three primary objectives:

- destruction of human pathogens through pasteurisation

- keeping the quality of the product without significant loss of flavor, appearance, physical and nutritive properties, and
- selective control of organisms which may produce unsatisfactory products

Milk processing plant procedures seek to:

- prevent further bacterial contamination of raw materials
- reduce bacterial numbers in milk
- protect the finished product from recontamination through careful handling, proper packaging and storage

Pasteurisation is the means whereby raw milk is rendered safe for human consumption. It is the process of heating milk to a sufficient temperature for a sufficient length of time to make it free of pathogens, however, not totally free of bacteria.

Bacteria in Milk

As stated earlier, certain organisms are capable of surviving pasteurisation and/or refrigeration processes. These bacteria are an important concern because they reduce product shelf-life. Those of major significance are:

Thermoduric Bacteria

- common in raw milk
- they survive pasteurisation and include:
  - Enterococci
  - Micrococci
  - Brevibacterium
  - Lactobacilli

Psychrotropic Bacteria

- common dairy product contaminants
- these grow at refrigeration temperatures

— they do not survive pasteurisation
— can produce off-flavours

Spore-formers

— common contaminants
— survive pasteurisation
  — Clostridial spp.
  — Bacillus spp.

The primary source of these bacteria is the environment: air, dust, dirty equipment, operators, etc. Therefore, proper cleaning and sanitising procedures are necessary for quality control. Grade A milk quality standards allow a maximum of 100,000 bacteria/ml. in raw bulk milk. Chronic offenders of these limits risk losing their license to sell milk to the Grade A market. Most dairies are able to maintain bacteria counts between 5 to 10,000 per ml. When high counts become a problem it is generally due to one or more of the following:

— improper cleaning of milking equipment (the most common cause of high bacteria counts in milk)
— improper cooling of milk
— occasionally, a herd experiencing a high prevalence of infection due to Strep ag. or Staph sp.

Somatic Cell Counts

Somatic cell counts represent another important milk quality parameter. As discussed earlier, milk with high somatic cell concentrations reduces cheese yields due to the lowered casein content. In addition, high cell count milk generally contains increased amounts of proteolytic and lipolytic enzymes (lipase). These presence of these enzymes in milk increases the potential for off- flavours and odours. Somatic means body and thus a somatic cell is a body cell. There are three types of somatic cells typically found in milk: epithelial cells, macrophages, and polymorphonuclear leukocytes (PMN). Cell types found in milk obtained from non-infected glands are predominantly epithelial cells and macrophages.

Milk from infected glands, however, generally contains high concentrations of PMN's with little or no increase in other cell types. Consequently, somatic cell counts are an important indicator of udder health. Somatic cell counts are made available to dairymen from a variety of sources including milk quality laboratories operated by state and local regulatory departments, dairy cooperatives, DHIA-SCC program, and veterinary diagnostic laboratories. In general, cell counts from herd bulk milk consistently in excess of 500,000/ml are indicative of a high prevalence of mastitis in the herd.

Mastitis causes a shift in the composition of milk. In addition to lowered amounts of casein, lactose and fat levels are decreased particularly in milk with somatic cell counts in excess of 2 million. Because the bacterial quality and somatic cell content of raw milk are important to product shelf-life, flavor and yields (particularly cheese), milk processors strive to obtain the highest quality raw product possible from their producers.

Antibiotic Residues

Antibiotic residues pose a significant public health threat. Consequently, milk in Florida is routinely monitored by dairy cooperatives and the Florida Department of Agriculture and Consumer Services, Division of Dairy Industry. The official test in current use is the Bacillus stearothermophilus disc assay. It is particularly sensitive for penicillin but can detect other inhibitors as well. The vast majority of antibiotic residues in milk occur by accident. Dairymen can avoid residue problems by:

- properly identifying treated cows
- informing milkers of the need to withhold and the method for withholding milk
- keeping an accurate record of dates and times of treatment
- following label directions and veterinarians' advice for withholding times

— having milk tested from suspect cows if uncertain about treatment or withholding time
— having tank milk tested when it is suspected of having milk containing antibiotic residue
— isolate purchased cows and test their milk for residue prior to their entry into the milking herd.

Dairy Cooperatives

Dairy cooperatives are organised by dairymen for the purpose of marketing milk. Thus, instead of buying milk direct from farms milk processors buy their raw product from the dairy cooperative. Dairymen are in turn paid by the cooperative for the milk they produce. Florida's dairies supply approximately 80% of the state's marketing demands. Therefore, at certain times of the year some milk must be imported to satisfy processing needs.

During other times of the year, milk production is in excess of market demand and milk must be exported out-of-state. These daily marketing difficulties could be both expensive and time-consuming problems for dairymen. Consequently, the majority of producers belong to milk marketing cooperatives. Cooperatives serve the dairymen by promoting dairy products, providing an effective lobby for political concerns, and informing members on a variety of dairy industry issues such as water quality, waste management, and milk pricing.

Dairy Farmers Incorporated (DFI), represents the members of Florida's two regional cooperatives:

1. Florida Dairy Farmers Association
2. Tampa Independent Dairy Farmers Association.

Regulatory Control of Milk Products

The milk sanitation program of the United States Public Health Service evolved for three reasons:

1. the U.S. Public Health Service promotes the consumption of milk for good nutrition

2. the potential for milkborne illnesses is a significant public health threat
3. reciprocal acceptance of milk and milk products between political jurisdictions

This program has been one of the most successful in accomplishing its objectives. In 1938, milkborne disease outbreaks constituted 25% of all disease outbreaks resulting from infected foods and contaminated water supplies. Today, less than 1% of such disease outbreaks can be linked to the consumption of milk and milk products.

The United States Public Health Service/Food and Drug Administration, divisions of the Department of Health and Human Services, have developed a statement of policy and regulations with regard to milk quality. This model regulation is known as the "Pasteurised Milk Ordinance of 1978" (PMO). This document contains (among other things) the milk quality standards recommended to states, counties and municipalities. Following adoption by state legislatures, policy and standards are established for that state which may be equal to but not less lenient than those outlined in the PMO. It is here where the legal authority for milk sanitation programs originates.

In Florida, the charge for enforcement of the state-wide Milk and Milk Products Law is vested with the Florida Department of Agriculture and Consumer Services, Division of Dairy Industry. This division is composed of three functional units:

1. Office of the Director
2. Bureau of Dairy Inspection
3. Bureau of Dairy Laboratories.

Within this organisational framework reside the responsibilities for policy making, licensing and inspecting of dairy farms and processing plants, milk sampling and analysis, and enforcement of the Florida Grade A Milk Programme.

Role of Veterinarian

A thorough understanding of milk quality is an essential component of the knowledge base needed for a veterinarian in order to evaluate, plan, implement and monitor a mastitis control program. This is best demonstrated by considering what constitutes a "mastitis problem herd." Problem herds are herds that suffer one or more of the following:

1. a herd with bulk milk containing high somatic cell counts
2. a herd with bulk milk containing high bacteria counts
3. a herd with bulk milk containing antibiotic residues
4. a herd with an increased number of clinical mastitis cases.

Three of four listed are milk quality problems and if not corrected can result in suspension from the Grade A milk market. This has very significant economic implications to the dairymen and demands the veterinarian's immediate attention. When a dairyman is "shut-off" it means he can no longer sell his milk until the problem is resolved. This represents an emergency situation.

Finding a rapid solution for such a predicament could mean the difference between losses of several thousands of dollars or minimal losses as a result of effectively managing the crisis. This is not the case with the herd which is sporadically experiencing increased numbers of clinical mastitis cases. While this represents an important concern for the dairymen it usually does not threaten his milk market. In short, the dairyman's desire to implement a mastitis control program may more likely come as a result of a milk quality problems and not a clinical mastitis problem.

Consumers, retail distributors, milk processors, milk cooperatives, regulatory officials, veterinarians, and dairymen all care about milk quality and with good reason. The future success of any business enterprise depends upon the "quality" of the product it produces. The American auto industry would certainly testify to that.

Proper Milking Procedures

Proper milking procedures are important for the prevention of mastitis and for insuring complete milk removal from the udder. Mastitis can decrease total milk production by 15 to 20%. To minimise loss and achieve maximum milk yield, a practical milking management scheme should be followed.

The term "milking management" includes care for the environment in which cows are housed or pastured. The dairy cow should have a clean, dry environment. This helps reduce the potential for mastitis and increases milking efficiency by reducing time and labour to clean udders before the milking process.

Movement of Cows

Movement of cows should be in a quiet, gentle manner. If cows are frightened or hurried, the milk letdown process may be disturbed. Therefore, rough handling of dairy cattle should be avoided.

Clinical Mastitis

Milking may begin with a check of all quarters for mastitis. It is acceptable to strip milk onto the floor in a milking parlour or flat barn. Any cows that show clinical mastitis should be examined and appropriate action taken. If fore milking is not done, visual checking for inflamed quarters is done by milkers and herd health people.

Dry Udders and Teats Preparation

The object of udder preparation is to ensure that clean, dry udders and teats are being milked. The federal government's pasteurised milk ordinance (PMO) states that a sanitizer must be applied before milking. This task may be accomplished by using an approved sanitizer injected in the floor-mounted cow washers or by using a hose and water with a sanitizer on the parlour. Single-service paper towels or washed and dried cloth towels may be used.

Predipping

Predipping with teat dip has become popular. The advantage may just be getting the water out of the milking barn so wet udders are not being milked. The procedure for predipping involves washing of teats with water and a sanitizer. The teats are then dried with an individual paper towel and dipped or sprayed with the sanitizer. A 30-second contact with sanitizer is needed to kill organisms. Then the sanitizer is wiped off of the teat with a paper towel.

The cows are milked and teats are dipped again with the same type of sanitizer to prevent chemical reactions that could cause irritation to teats. Predipping may be beneficial in reducing mastitis, but the actual dipping, dip contact time, and wiping with a towel increase the total milking time. If the dip is not wiped off, excessive chemical residues in milk may occur. If contact time is not sufficient, then it's a very expensive premilking regime.

Attach the Milking Unit

To attach the milking unit to the teats, apply the cluster allowing a minimum of air admission and adjust to prevent liner slip. Air entering the unit may cause the propulsion of mastitis organisms from one infected teat into a noninfected teat. This also may happen when one teat cup is removed before the others.

Machine stripping usually is not needed on dairy cows. Machine stripping should not take more than one minute and no air should be allowed to enter the teat cups while this is being done. A downward force applied to the cluster while massaging the udder with the other hand is all that is needed. Following milk-out, the machine should be removed only after the vacuum to the teat is shut off. This is accomplished most commonly by use of a vacuum shut off valve or milk hose clamp which prevents the backjetting of bacteria from one teat to another.

Backflushers

Backflushers have been developed to sanitize the liners and claws between milkings. Most units on the market have four or five cycles. The first cycle is a water rinse, followed by an iodine or similar sanitizer rinse, a clear water rinse, and positive air dry cycle. Research has demonstrated that backflushers do reduce the number of bacteria on the liners between cows, but do not reduce the number of bacteria on teats.

Backflushers also may stop the spread of contagious organisms, but this can also be accomplished at a much lower cost by teat dipping. There is no effect on environmental pathogens that are encountered between milkings. Backflushers may be effective in stopping the spread of contagious mastitis; however, there is limited research to support this view. Because of the high initial cost, the need for daily maintenance, and limited efficacy, backflushers are not routinely recommended.

Post-milking Teat Dipping

There is only one way to effectively stop the spread of mastitis in the dairy herd, and that is by applying teat dip to every quarter of every cow after every milking. Teat dips are used to remove milk residue left on the teat and kill organisms on the teat at the time of dipping. They also leave a residual film of sanitizer between milkings. Teat dips have been shown to effectively reduce mastitis caused by S. aureus and S. agalactia, the most common types of mastitis found in Florida.

Post-milking teat dipping is effective in eliminating environmental organisms E. coli and Strep. uberis on the teats after milking. These pathogens are found in the cow's surroundings; if there is udder-deep mud, the teat dip will be removed and a new infection may occur.

Types of Teat Dips

There are many effective teat dips, including iodine at 0.1%, 0.5%, and 1.0%. Also, although it is not labelled for teat dipping, hypochlorite at 4.0% with a sodium hydroxide content less than 0.05% was effective in field trials. There are many more teat dips on the market that are effective in preventing new infections. Effective coverage of the teats is more important than the type of dip being used.

Dip or Spray

If contagious bacteria is present in your herd (Strep. ag., Strep. dysgalactiae, Staph. aureus, or mycoplasma), you must dip the whole teat to the base of the udder to stop the spread. Wand sprayers are acceptable for herds that have environmental mastitis, since teat colonisation is not a factor. Hand-held spray bottles are almost worthless in getting proper coverage of dip on the cow's teats, so they should not be used.

Requirements of a Milking System

Amount of Vacuum

The amount of vacuum used to operate the milking system is quite small, less than 2 CFM per unit. Extra CFMs are needed to compensate for vacuum losses that occur naturally in the system: head loss and resistance of pipes, elbows, etc. Thirty CFM ASME are needed in any system just to keep the system operating. Leaks account for a 10% loss (the difference between pump CFMs and system CFMs). Milk meters, unit slippage, air leakage used when applying the unit, and unit fall-off also consume varying amounts of vacuum depending on the skills of the operators.

Stated vacuum requirements vary greatly and are based on personal bias rather than research. The usual method is expressing CFMs per unit, which usually undersizes small systems and greatly oversizes large systems. Electric motors that run vacuum pumps are usually either 5, 7.5, 10, 15, or 20 HP and 1 HP motors will deliver 10 CFM ASME on an oil pump, or 7.5 CFM on a water pump. You must determine the pump size for your vacuum usage range.

For determining the size of vacuum pump needed, include 30 CFM for running the system, 10 percent loss due to leaks, and 3 CFM per unit for each unit (2 CFM to milk and 1 CFM added for milk meters and other losses). This should provide more than adequate vacuum for most systems. This does not include extra capacity for one half of the units open on the floor. A milking system should not be designed for poor milking procedures. Adequate milking vacuum level to prevent liner slip will reduce fall-offs and most automatic take offs will shut off automatically on fall-offs.

Milk Line Size

Research has indicated that milk lines have been oversized in the past. Oversized lines are expensive to purchase and expensive to clean, because they require more hot water and chemicals.

Pulsator Line Size

A 3-inch diameter plastic pulsator line is used for rigidity and ease of taping for pulsators. There is no need for a size larger than 3 inches.

Vacuum Controllers

Type: Diaphragm type are the most common and sensitive—Sentinel, Westfalia, Delaval Servo, DEC Servo, etc. Dead weight controllers should be avoided.

Location: In as clean an environment as possible and also easily accessible for cleaning. The vacuum controller should be located where the manufacturer recommends, usually between pump and the trap. If located on the balance or reserve tank, the diaphragm type should be double elbowed off the tank.

Maintenance: Clean as often as recommended or at least once a month. In systems with excessive amounts of pump capacity, cleaning must be carried out more often.

Pulsators

A pulsation rate of 40-60 pulsations per minute is adequate.

Pulsation ratios of between 50:50 and 70:30 are adequate. A 70:30 ratio with 60 pulsations per minute gives the fastest milking. Alternate and simultaneous types are both acceptable. The choice of pulsation rates, ratios and type are more a matter of personal choice than based on any scientific data. Those commonly available from manufacturers work well. Pulsators should be cleaned monthly. Dust caps more often if dust, moisture and insects are common in the area.

Line Vacuum Level

Vacuum levels used are more based on choice than fact, as any vacuum between 12" Hg and 15" Hg may be adequate. Most high lines milk better at 15", low lines may work at 13.5" Hg. For maximum milking speed milk at 15" vacuum, 60 pulsations per minute, 70:30 ratio. But if "liner slip" or "fall off" are a problem, raising the vacuum to 15" usually helps. Vacuum levels above 15" Hg should be avoided. Line vacuum should be checked at the pump, the receiving jar, and by the controller.

Differences in vacuum levels between pump and receiver should be less than 0.5"Hg. Higher readings, indicating greater pressure drops, relate to decreased CFM at receiver. Greater pressure drops are influenced by small line sizes, excessive elbows, or unreasonably high air flows. Differences in vacuum levels between receiver and controller should be less than 0.2" Hg. Higher readings indicate higher vacuum differences which influence controller performance.

Role of Milking Machine

The role of the milking machine in causing mastitis has been debated for years. The most recent research has shown that a milking system properly installed, operated and maintained has little effect on mastitis. Research has demonstrated that cyclic vacuum fluctuation (that occurs normally in the milking cycle) and irregular vacuum fluctuation (that occurs when units fall off or are carelessly handled) occurring simultaneously will increase the new infection rate.

Line slip during milking will also increase the infection rate. Infection is increased because small droplets of milk are back-jetted against the teat end, with some organisms being forced into the teat. This is more of a problem with the way the unit is operated rather than with the milking machine itself. The most common way that the milking machine influences mastitis is that the system is not maintained properly and the milking machine may damage the teat end, thus increasing the chances of mastitis organisms entering the udder.

A vacuum level of above 15" Hg. will damage the teat end. This is caused by a malfunctioning or dirty vacuum controller. Malfunctioning vacuum controllers can also increase vacuum fluctuations. If the pulsators are dirty and no air is admitted, there will be a very short or no massage cycle which can cause damage to the teat end.

A recent survey in Florida indicated that 75% of the farms surveyed had pulsators malfunctioning and over 50% of the dairies surveyed had poor vacuum controller response. This survey indicates that these two important parts of the milking system are not being maintained. They must be for proper udder health! This fact sheet in conjunction with fact sheet DS-5, Advanced Milking Equipment Analysis Data Form, can give a complete analysis of the milking system. It also explains which parts of the system should be checked and how to do it. DS-5 also provides a record of this analysis for future reference.

Vacuum Pump

Evaluate every two months.

- Location - Pumps should be located in a well ventilated area, clean and dust free, as close to the parlour or milking barn as possible, but isolated enough to keep the noise level in the milking area to a minimum.
- Check belts for wear and tightness.
- Check rubber hoses for holes and hose clamp tightness.
- Check filters (if present) for cleanliness.

— CFM of pump should be checked with an orifice flow meter. The pump should be warm when checked. Do the check as close to the pump as possible, CFM is always checked at 15" vacuum, and recorded that way. Pump CFM's may also be checked at milking vacuum level if different than 15".

— Pump should function within 10% of factory rated capacity, if not, should be cleaned or rebuilt.

System Vacuum Level

Check several places in system. Resident gauge reading should also be checked: high lines 15", low lines 13.5" Hg.

System Check (Leaks)

Reconnect vacuum pump to the system, remove or shut off controller, shut off pulsators, shut off claws or crimp milk hoses and tape them. If possible remove top of receiver jar and insert the flow meter into receiver top adapter, start pump and record CFM's—leaks should not be more than 10% of pump capacity, but often are. If receiver jar is such that it can't be used, place flow meter in pipe where controller was removed. If leaks account for more than 10% of total capacity check pipe joint leaks, etc. Minimum system CFM's are 30 CFM (A.S.M.E.) + 1.5 CFM per unit.

Controller Responses

Reconnect system into milk mode, units not on, claw still off or hoses taped, controller in place and operating. Insert flow meter into receiver jar inlet or top, close flow meter, note vacuum level, let in air through flow meter in 10 CFM increments. Note vacuum: there should not be a vacuum change + or -0.5" vacuum up to 75% of capacity of system check CFM's. If controller response is poor, check for dirty filter(s) or dirty controller body. Most dead weight controllers will not pass this test, diaphragm type are most responsive.

Vacuum Controller Leakage (optional)

Remove or block off controller, place flow meter in receiver jar with system in milk mode (as in controller response check above) obtain CFM's at 0.5" Hg or 2 kpa below normal operating vacuum level, record CFM's. Reconnect vacuum controller with flow meter in same place, record CFM's at 0.5" Hg below vacuum level. The difference between the two should not be more than 10 CFM. Some diaphragm regulators use up to 10 CFM as a bypass procedure normally. Non-diaphragm regulators should not lose more than 1 CFM.

Pulsators

Should be inspected visually for cleanliness, air inlets should be clean, free of dust and spider webs. Pulsators used in flat barns should be checked for dented air inlets (where they are allowed to hit the floor). A graph should be made of each pulsator function while all other pulsators are operating on the system.

In flat barns both sides of the barn should be checked. Pulsation rate and ratio should be recorded and any abnormalities observed and recorded. If you do not make a graph a western dairy meter may be used. If nothing else is available, a vacuum gauge may be used to check if pulsator opens and closes—milking vacuum to 0" Hg.

Cluster

Pulsator hoses and short air hoses should be checked for leaks, holes, cracking, etc.

- Check claw vent: it should be open and clean if no liner vent is present, should be closed if liners are vented. If liner vents are used they should be clean and the claw vent sealed.
- A vacuum shut off should be present on the claw or on the milk hose before the claw.
- Liners or inflations should be free of holes and liners changed regularly; 1200 cow milkings for moulded liners and 600 cow milkings for stretch bore liners.

Automatic Take Offs: Sensor jars should be located at udder level or lower if possible. If floats are used they should be right side up. Older electronic sensing tubes should be checked for old collapsed rubber tubing.

Milk Line: Slope at least 1" per 10' (1 ½" per 10' is better) of line and should be looped, all inlets or nipples should be at the top of the line.

Vacuum Supply Line: If of P.V.C. it should be installed with enough supports to keep it from sagging and designed to be taken apart and washed.

Pulsator Lines: Should be supported to prevent low spots and have ports so it can be washed. Preferably looped and installed low enough that the pulsators can be reached without a ladder.

Stray Voltage: Anything above 0.5 volts A.C. should be avoided. A sensitive volt meter must be used and must not pick up D.C. voltage on the A.C. scale.

Places to Check

— Milk line to claw, one lead to the claw - the other to the milk line.

— Claw to the floor, one lead to the claw - the other to the floor.

— Bulk tank outlet to the milkhouse drain.

Wash System

— Hot water: temperature 160° preferable.

— Air injector must work for proper cleaning, a loud hiss at regular intervals usually means it works. Filter should be clean if present.

References

Beetz, A., "Grass-based and seasonal dairying", *ATTRA*, Fayetteville, Ark., Dec. 1998.

Booth, G., "It's far deeper than machinery: Intensive rotational grazing on the Molitor Farm", *Greenbook 1999. Energy and Sustainable*

Agriculture Program, Minnesota Department of Agriculture, St. Paul, Minn., pp. 1999.

Hattori, N, Tobioka, H, Kinugawa, K, Fujiwara, N, Yamamoto, T, Nishi, R., "Preliminary application of nursing facilities for calves separated from grazing cows: Suckling milk yeild, composition and body weight of calf", *Proceedings of Faculty of Agriculture - Kyushu Tokai University,* 1997.

Shirley, C., "Less milk, more profit: Organic feed and rotational grazing keep this dairy green", *The New Farm;* Sept/Oct. 1991.

7

Milk Processing Methods

In developing countries, it is quite common to find milk production areas at great distances from areas of heavy demand for dairy products. These areas of demand are usually concentrated in the capitals and major cities. In an attempt to meet this demand in part, developing countries have tried to establish dairy industries near urban consumption centres. While there is overproduction in some parts of the country due to lack of outlets, dairy plants often work at only 20 percent of their capacity because of the long distances between the milk producing zones and the urban consumption centres where the plants are situated.

Costs of milk collection for plants located near the capital are very high due to fuel prices, poor road conditions (particularly during the rainy season), the cost of spare parts and the mileage required to be covered. For economic reasons, therefore, dairy plants have tried to restrict milk collection to nearby milk-producing zones, rounding out their daily output through the importation of dried milk and butter oil used for reconstituted milk. The cost of milk collection represents approximately 30 percent of the processing cost of the finished product (packaged pasteurized milk).

The traditional demand in developing countries is for fresh milk in the cities and fermented milk in the countryside. The dairy plant therefore primarily produces fresh and reconstituted milk, pasteurized, and packaged in plastic bags. Governments generally consider milk a basic commodity and consumer prices are strictly controlled. Plant profits from milk sales are usually quite low, forcing the management to market another, more remunerative product not subject to price control. A frequent choice is yoghurt, which is sold to wealthier people and to expatriates.

Besides the market for local milk products, there is also a market for imported products which, in order of importance, are dried milk for babies, tinned milk and cheeses. As a result, the problem in many developing countries is that while the small milk producer has no regular outlet for his milk, consumers pay high prices for imported dairy products.

Dairy Processing Plant in Village

First of all, a dairy processing plant should offer producers a guaranteed daily outlet for their milk supply. During the dry season, when domestic milk production is low, milk-pedlars are a common sight buying milk from producers who are far from urban consumption areas. However, during the rainy season, due to poor road conditions and increased milk production, this milk is not collected. If a milk producer were guaranteed a regular, fair income, it would provide the incentive for him to move from a position of self-sufficiency in milk for the family into one of selling, thus boosting milk production in his area.

Remote milk-producing areas are often good livestock production regions, but because of lack of commercial outlets they do not receive the necessary incentives for boosting milk production. The introduction of a dairy industry, modelled on European systems, into a developing country has only very rarely involved the producers themselves. And even when it has, usually only the big milk producers located near large towns have been involved.

The introduction of a processing plant or unit, at village level, concerns principally the local producers themselves. The size and simplicity of the unit should allow participation by small producers who are aware that self-help is a more immediate prospect than Government or bilateral assistance.

Village Milk Processing Unit

- by forming associations, the small producers, who live directly on the meagre resources brought to them by their cattle, can set up the unit. Many big herds belong to "entrepreneurs" in cities who entrust their stock to the care of herdsmen, paying them only the equivalent of one day's milking a week.
- by presenting a type of project which suits their needs— simple, practical and which allows them to work together within their own environment.

Milk-producing Area

- The site should be in a remote, inaccesible, traditional milk-producing area. Such areas, where communications are difficult and cattle-raising is a normal component of the family livelihood, are very common in Latin America, Africa, the Near East and Asia.
- An area should be chosen where milk collection from the capital is not possible.
- In an area where cattle, and milk in particular, should be the main regular source of cash for purchasing the family's domestic needs such as clothing, sugar, etc.
- In an area where there is plenty of water to allow hygienic processing of dairy products which requires an average of five litres of water for each litre of milk processed.
- In an area where the quanity of milk to be processed is available within a ten-mile radius since the time required for the transport of the milk should not exceed three hours. In tropical countries this is generally recognised as the upper limit beyond which milk cannot be pasteurized or heat-treated.

Locally-produced Dairy Products

- produces dairy products for which there is demand in the cities and in the villages;

— produces dairy products not supplied by dairy plants located near towns;
— produces dairy products which:
  — can easily compete with imported products;
  — do not require costly or sophisticated equipment;
  — require neither an expensive source of power nor a particularly complex infrastructure;
  — can be easily and cheaply moved without lowering the quality of the finished product;
  — can be sold in small unit sizes so as to reach a large number of consumers.

Locally-produced dairy products, particularly those such as cheeses, butter and fermented products, are frequently considered by the more affluent consumers in the capital cities of developing countries as second-rate compared to imported products.

In fact, locally processed milk products are often considered even by the producers themselves as merely by-products, surplus to their family needs. The project needs to demonstrate that with appropriate technology the finished products can be just as good as, and often better, than imported products considering the time required to clear imports through customs. At a minimal cost their presentation can also be as good.

Village dairy unit should be:

— as simple as possible;
— as clean as possible;
— as cheap as possible;
— as profitable as possible.

The village dairy unit is a tool:

— for a group, association or cooperative,
— for pooling resources so as to exploit their milk surpluses,
— to enable people to organise, and obtain the resources they need for improving their welfare.

General Model of Milk Processing Unit

Milk-producing Area

The unit should be located as centrally as possible within a given milk-producing area, near a source of water, or in a place where water is available. The site should be cool and well-ventilated. Sometimes not all these conditions can be met. The most important factor is availability of water. It should be remembered that on average five litres of water are required to process one litre of milk.

Building

Milk and dairy products are biologically active substances which are influenced by their environment. Cheese quality and reliability depend largely on the surroundings in which cheese is manufactured. An unused building can be purchased or leased and adapted for milk processing operations, or a new building can be constructed. It is not uncommon to find in some remote milk-producing areas, abandoned milk collection centres which may be suitable.

For a new building, the following factors should be taken into consideration:

— the walls should be built of local stone and the inner walls lined with a lime-cement mixture for easy cleaning;

— the cement floor should have a 2 to 3 percent slope for draining water used in cleaning;

— windows should be sufficient to provide adequate ventilation.

For village cheese manufacture, the second component of the building is the ripening cellar.

The most important features of a ripening cellar are maximum moisture (80 percent relative humidity) and low temperature (8 to 12°C). To achieve or approach these standards, a room partly below ground level is recommended. It should be about 2.5 m high. The floor of the room should be dug to a level of some 1.5 m below ground, with windows or openings made in the upper walls to lower the temperature of the cellar by a circulating draught, particularly important during the night.

Building size will of course depend on the quantity of milk received during the peak production period. An average quantity of milk which can be processed by a small-scale unit amounts to 100 to 500 litres per day. For these quantities the building area should be some 50 sq. m.

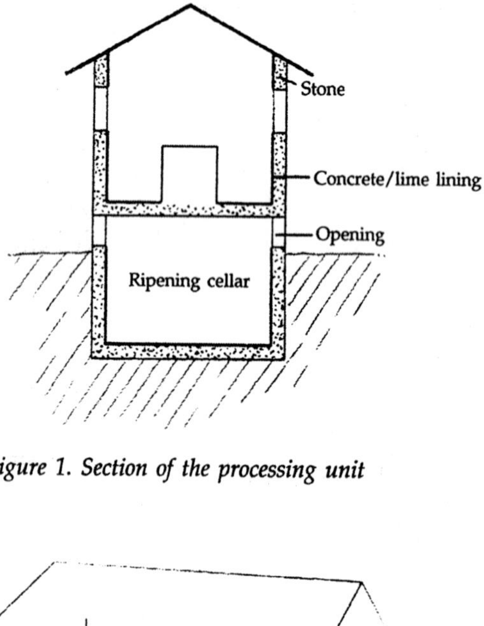

Figure 1. Section of the processing unit

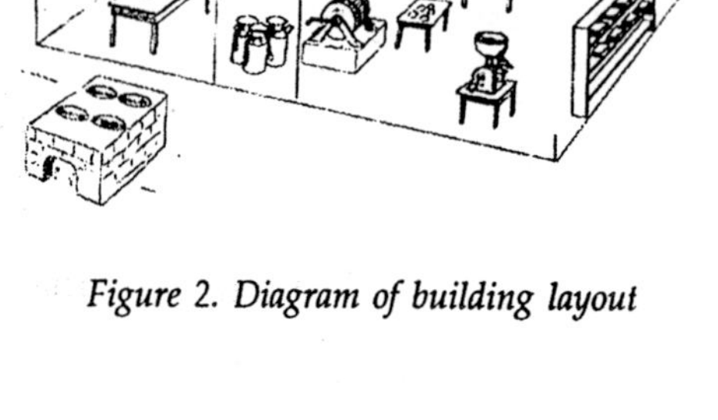

Figure 2. Diagram of building layout

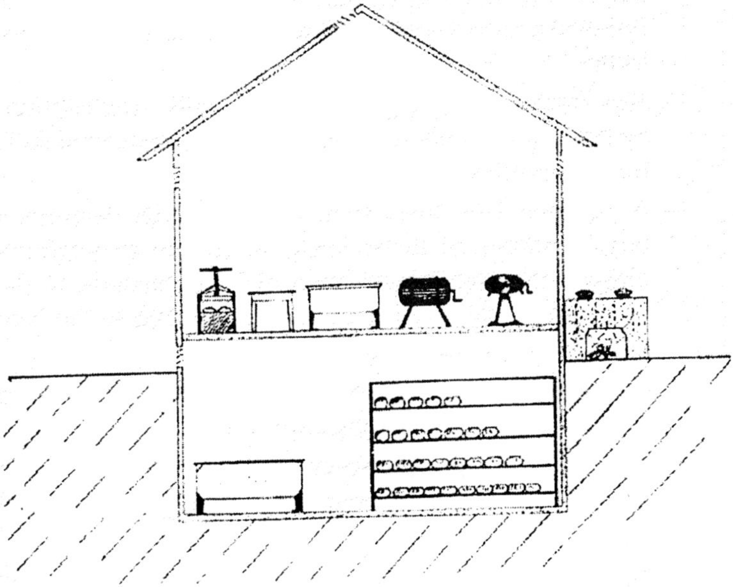

Figure 3. Sectional diagram of processing unit and equipment

Milk Processing Operations

As previously pointed out, milk processing operations will take place far from urban consumption centres. In these areas the quantities of milk hardly exceed an average of 500 litres per day. The products made must be able to withstand long periods of transportation, often under difficult climatic conditions.

Butter, cheese and processed cheese may require lengthy transportation, given the distance from consumer centres, whereas buttermilk and yoghurt can be marketed in the vicinity of the processing unit. The whey will be returned to the dairy farmers.

The main steps in obtaining the above products are:

— *Standardisation*: Standardisation is an operation producing milk with a constant butterfat content through partial, manual skimming. The operation makes it possible to standardise the composition of the finished product and to set aside part of the cream for butter.

— *Heat treatment*: Pathogenic germs in milk are destroyed by heating the milk to a minimum temperature of 63°C for 30 minutes.

— *Inoculation*: Due to heat treatment, which destroys a large number of lactic bacteria, cheese or yoghurt-making requires the addition of lactic bacteria to the milk. These bacteria are selected according to the type of finished product required.

— *Clotting*: Milk changes from a liquid to a solid state through the use of a coagulant: rennet.

— *Curd-Separation*: In cheese-making, the milk after coagulation is cut and separated into a liquid whey, and cheese curd.

— *Ripening*: This phase of cheese-making allows cheese texture to become homogeneous and the aroma to develop.

— *Churning*: In this operation, cream is churned to produce a semi-solid product which becomes butter.

— *Melting and emulsification*: Defective cheeses are melted and emulsified with salts to obtain a solid consistency and, after cooling forms processed cheese.

In addition to the eight steps mentioned above, milk collection, milk analyses and the marketing of the finished products should be equally regarded as important operations.

Equipment Needed

The equipment needed to run the dairy processing plant depends on several factors: how much milk is to be collected, how far and how scattered are the milk producers, what kind of product is to be produced?

In the standard milk processing pattern, commencing with milk collection and ending with the sale of the dairy products, the following equipment would normally be required:

For Collection

— Plastic milking pails are often an improvement on the utensils commonly used.

— The producer should use small aluminium milk cans of 5-10 litre capacity for transporting the milk, while the collector should use 30-50 litre cans.

— If the milk needs to be collected, the following will be needed: the use of a bicycle, a 50 litre milk can, a graduated cylinder lacto-densitometer and a measuring pail.

The amount and density of the milk collected from each producer should be entered in a notebook. The following is a sample entry:

No.	Name of producer	Amount delivered (litres)	Density	Signature
1.	Ram Singh	10	1.030	
2.	Iqbal	5	1.029	
3.	Charanjeet	15	1.038	
4.	Subhash	12	1.030	
5.	Chote Lal	5	1.032	
TOTAL:		47		

For Processing

— *Reception*: The following equipment is needed for the reception of milk brought in by producers themselves and by the collector: a milk scale and a pail.

— *Storage*: A milk funnel and 50 litre milk cans.

— *Standardisation/Cream separation*: A manual cream separator has to be used to skim a portion of the milk received.

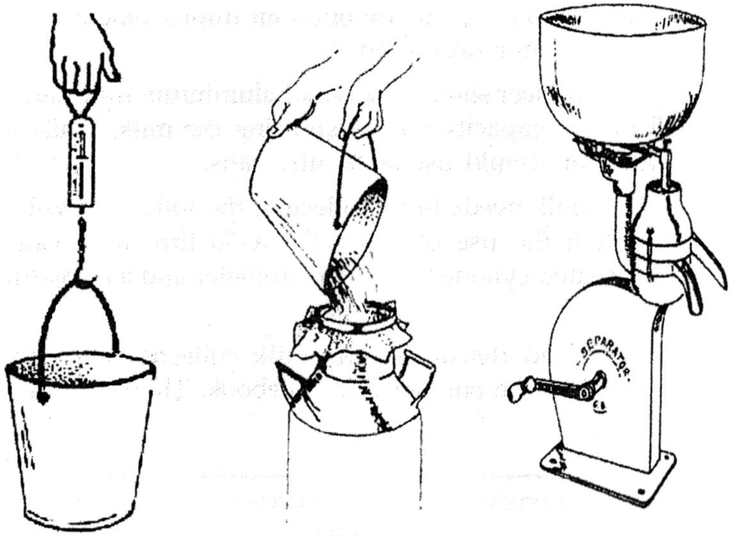

Figure 4. Milk scale and pail, Milk funnels filter and milk can, Standardisation hand-operated separator.

Heat Treatment

There are several possibilities for heat treatment, depending on the available power source. Under the least favourable circumstances, the sole available energy source is wood or peat. The best thing to use in this case is a "boiler/water bath" as in the model below:

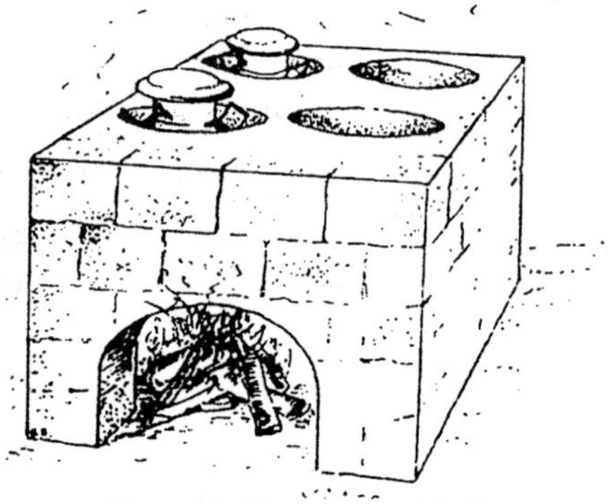

Figure 5. Cement-block boiler

More elaborate models can be built for wood or gas-fuelled heating (where bottled gas is available).

Figure 6. Wood-fuelled metal boiler *Gas-fuelled metal boiler*

For 100-500 litre quantities of milk, milk pasteurization with a plate pasteurizer is not recommended.

— *Cooling*: The milk is cooled with running water in a vat:

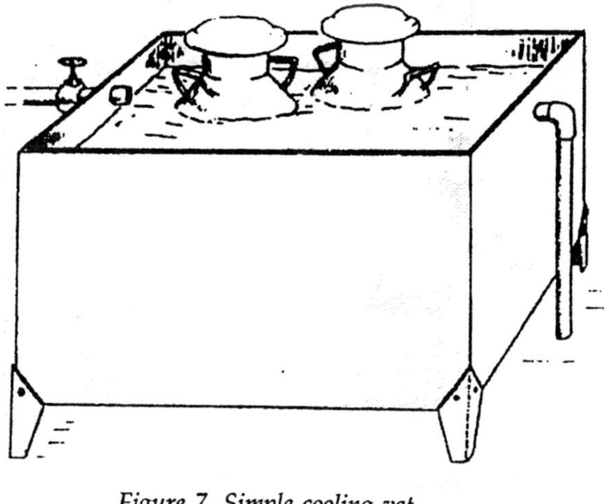

Figure 7. Simple cooling vat

or by water circulating in a jacketed vat.

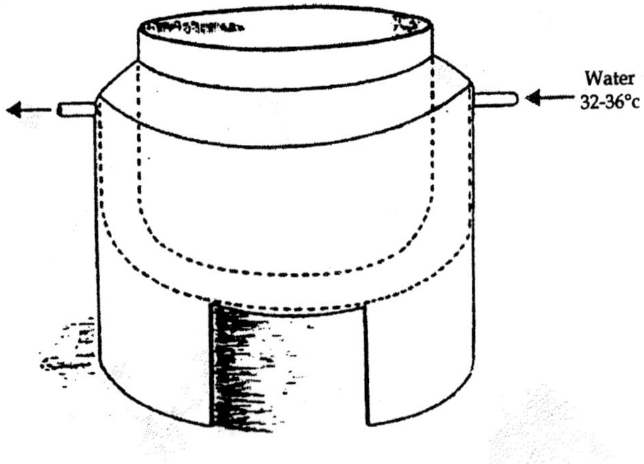

Figure 8. Jacketed cooling vat

— *Clotting*: The cheese vat can be of aluminium with a tap for draining the whey.

Figure 9. Simple milk clotting vat

A jacketed vat can also be used for milk clotting.

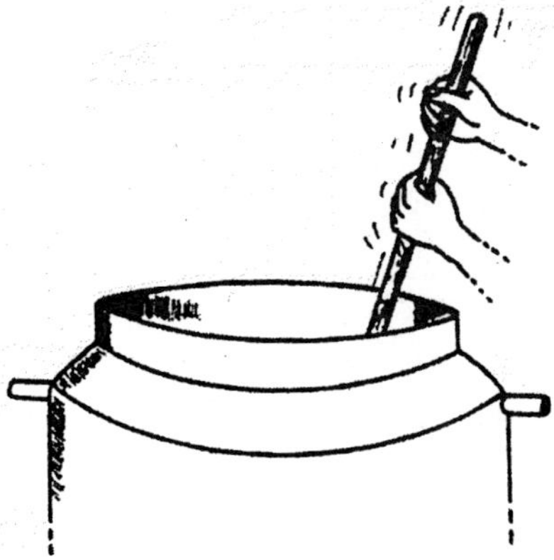

Figure 10. Jacketed clotting vat

There are other possible types of cheese vats, depending on the resources available to dairy producers. The following are a few examples:

— simple vat to process 50-100 litres of milk

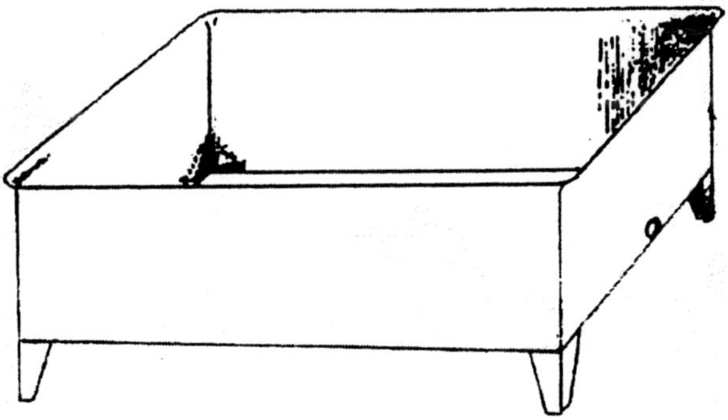

Figure 11. Low-walled clotting vat

— a more elaborate vat for processing 100-300 litres of milk: milk cooled by cold water circulating within its jacket:

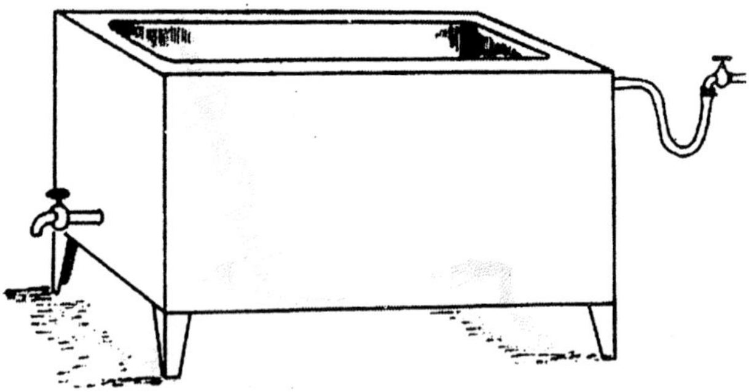

Figure 12. Clotting vat equipped for cooling

— for larger amounts of milk up to 500 litres, a vat equipped for heat treatment, cooling and clotting can be designed as shown in the following diagram:

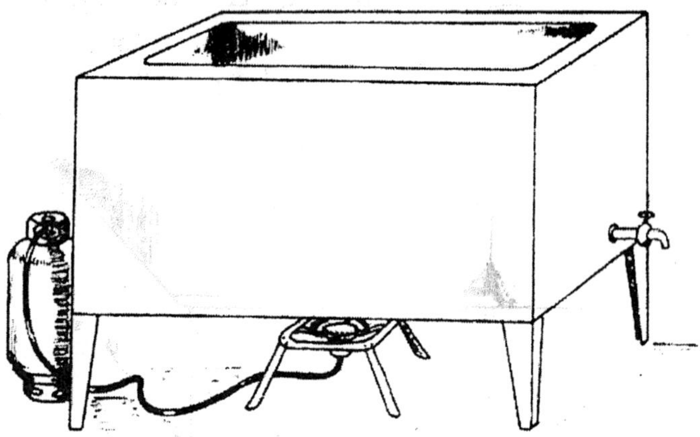

Figure 13. Multi-purpose vat

— *Draining*: The cheese is drained in moulds set on a slanting surface to drain the whey. The moulds, which give the cheeses their characteristic shapes, vary greatly in size and form.

The simplest way to make cheese moulds is to cut a plastic pipe generally used for drainage into 10 cm sections.

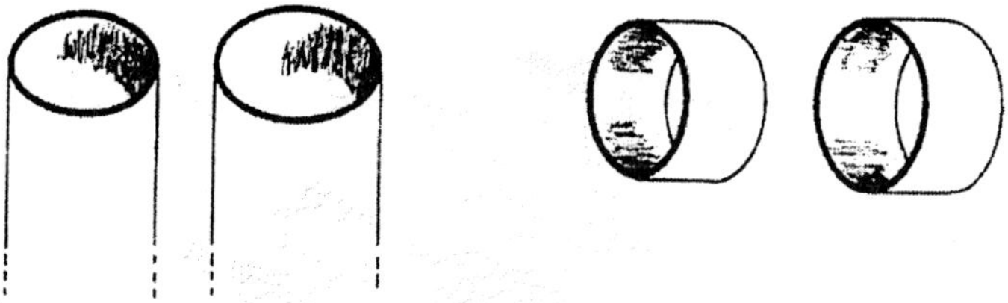

Figure 14.

The resulting cylinders are perforated as in the model below:

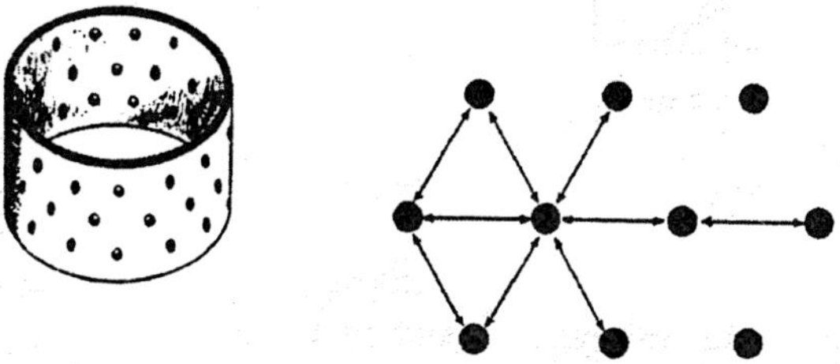

Figure 15.

Bases and Lids are of wooden discs slightly smaller in diameter than the cylinders.

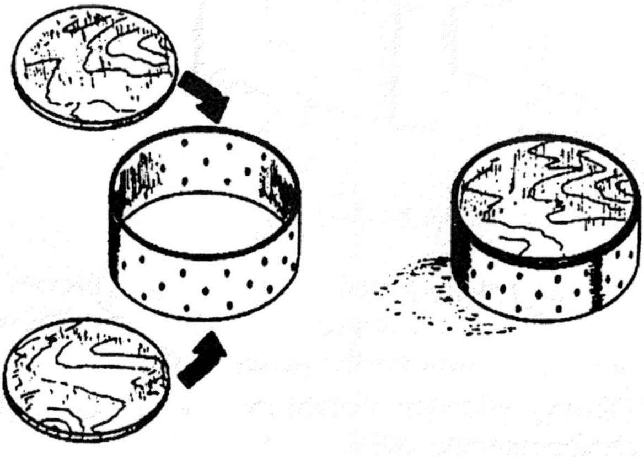

Figure 16. Moulds with bases and Lids

Cheese moulds may also be made of wood, their shape varying according to the type of cheese made.

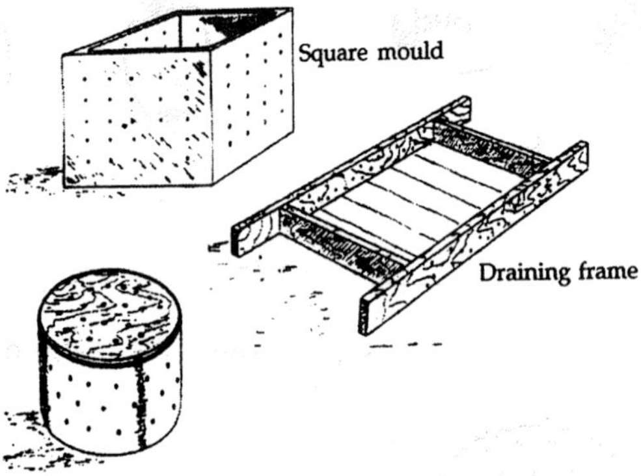

Figure 17.

The cheese-draining table, slanted forwards to facilitate draining of the whey, is made of wood.

Figure 18. Wooden cheese-draining table

— *Pressing*: Different kinds of cheese (curd) presses can be made. The simplest press is made by placing weights or cement block on the moulded cheese.

— *Ripening*: Wooden shelves must be assembled for the cheese-ripening cellar.

Figure 19. Ripening shelves

For Marketing

Marketing, as previously mentioned, is a very important aspect of overall operations. Quality of presentation is important. Wooden boxes made to accommodate the specific cheese shapes are used to transport the cheese to consumer areas.

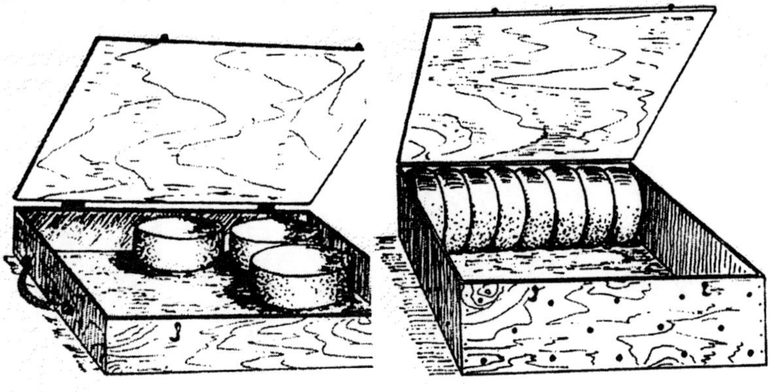

Figure 20. Wooden cheese boxes

Additional Equipment

An important component in a dairy processing unit, is the laboratory.

Laboratory equipment should include the following:

- To measure milk density:
  - Two or three lacto-densito-meters with glass cylinders
- To test milk acidity:
  - A Dornic acidimeter and the accessories.

Salut acidimeter used for selective testing of milk acidity immediately after reception of farmers' milk:

- Fat content of milk
- Preparation of cultures

To prepare mesophilic cultures for cheese-making or thermophilic cultures for yoghurt-making, a strain of starter culture is initially required. Culturing and sub-culturing is done in individual 5, 10 and 15 litre containers

References

Atkins, P.J., "The pasteurization England: the science, culture and health implications of milk processing, 1900-1950", In: Smith, D. & Phillips, *Journal of Food, science and regulation in the 20th century*, Routledge, 2000.

Gegner, L., "Value-Added Dairy Options", *ATTRA* (Appropriate Technology Transfer for Rural Areas), Fayetteville, Ark., Aug. 2001.

8

Organisation and Operation of the Milk Processing Unit

A village milk processing unit usually involves a group of milk producers living within a given area near the unit. It is therefore reasonable that the majority of the producers will deliver their own milk in the morning direct to the processing unit. The milk from the evening milking is retained by the family for home consumption.

For producers who live further away or small farmers who do not wish to make the trip with a small quantity of milk, collection can be arranged within a radius of no more than 10 kilometres. A milk collector on a bicycle picks up the milk at collection points, which are simple shelters within which the milk can be kept in the shade.

At the milk collection point, the milk collector tests milk density with the lactodensitometer, tests milk acidity, and measures milk volume using a measuring can. The milk is then filtered and poured into the milk can.

The entire milk collection round should take no longer than two hours. The milk collector has a little notebook in which he enters the volume and density of the milk delivered by the producer. Payments for the milk are made twice a month. The person responsible for milk collection may either be paid by the village cheese unit or by the producers themselves. For purposes of simplification, the second choice is preferable.

Reception of the Milk

Reception of the milk delivered by the farmers themselves or by the collector should take place in the very early morning. The producers' milk is weighed, density checked, the milk filtered, and then poured into milk pails.

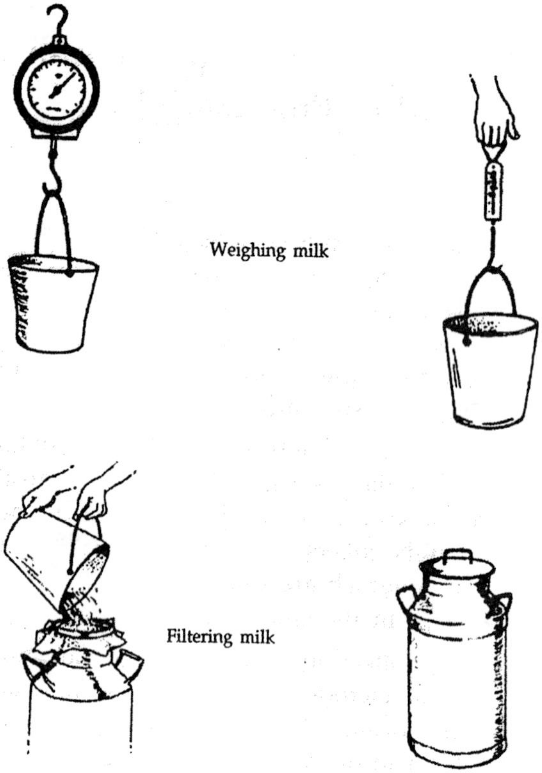

Figure 1

Milk density is measured as shown in the diagram below:

- The producer's milk sample is poured slowly into a test-tube to avoid foaming;
- The lactodensitometer is introduced into the test-tube and, once the level is steady, the number is read:

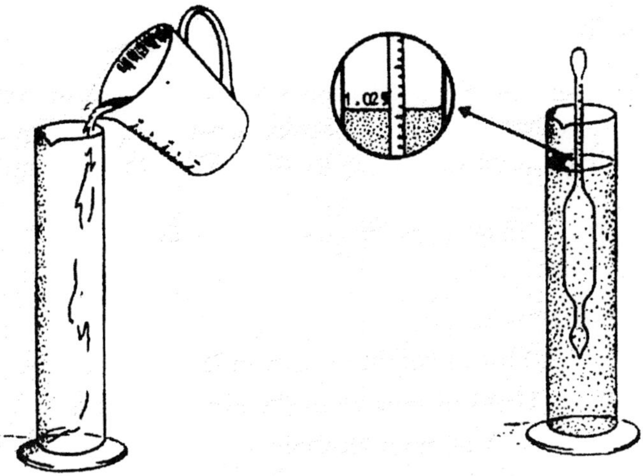

Figure 2. Interpretation of readings

For cow's milk an approximate estimation of density can be read as follows:

Lactodensitometer reading	Result
1.028 to 1.033	Normal milk
less than 1.028	Diluted milk
1.033 to 1.037	Skimmed milk

Standardisation of Milk

When milk from all the producers has been poured into the milk cans, a sample is taken from each can and mixed together to obtain an average sample. This sample is tested for milk acidity and fat content.

Milk Acidity

— Put 10 ml of the milk sample into a glass.

— Add 3 to 4 drops of phenolphthalein.

— With the dropper, add the NaOH $\frac{N}{9}$ solution drop-by-drop into the glass until a stable pink colour results.
— Read on the graduated column the number of ml used. This gives the acidity of the milk in Dornic degrees.

$$15\ ml\ of\ NaOH\ \frac{N}{9} = 15\ Dornic\ Degrees$$

Fat Content

1. Put 10 ml of sulphuric acid in the butyrometer.
2. Add 11 ml of milk from the average sample.
3. Add 1 ml of amyl alcohol.
4. Shake the butyrometer to dissolve the milk elements.
5. Put the butyrometers in the centrifuge. Centrifuging should continue five minutes.
6. Next plunge the butyrometers vertically, cork down, into a water bath, temperature 65°-70° C, and leave them for five minutes.
7. The butyrometer, cork down, should be perfectly vertical when taking the reading at eye level. Read the graduation mark at the base of the meniscus, i.e. at the base of the curved upper surface of the fat column. In the example that follows, the degree reading is 3.6. The milk fat content then is 3.6 percent, or 36 g of fat per litre of milk.

Heat-treatment of Milk

Heat-treatment of milk is a most important factor in the quality of the finished product. After standardisation, milk must be heat-treated. This means bringing it to a minimum temperature of 63°C for 30 minutes. The various equipment suggested for heat-treatment can bring the temperature up to 63°C from between 40 to 60 minutes. The milk must be constantly stirred to maintain a homogeneous temperature throughout the heating and heat-treatment stages. The milk stirrer and the thermometer are important items of equipment.

Cooling Milk

When the milk has been kept at a temperature of 63°C for 30 minutes, it is then cooled down to bring it to a suitable temperature for cheese-making (approximately 35°C). The milk is cooled either by immersing the milk cans in a tank with cold running water, or by running cold water through the double sides of the cheese vat. During the cooling, as during the heating period, the milk must be stirred constantly.

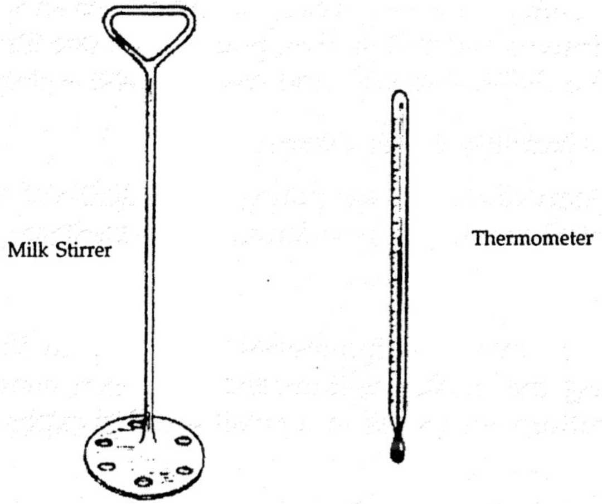

Figure 3.

Bearing in mind heat losses, cooling water should be shut off before the desired milk temperature is reached. For instance, if the milk temperature for cheese-making is 32°C, cooling water must be shut off when the milk temperature has reached 35°C.

Processing Aids

Preparation of Starter Cultures

One person only must be responsible for the preparation of starter cultures.

The simple method for preparing starter cultures is illustrated.

— The preparation of the starter cultures requires suitable strains of bacteria and a milk of good bacteriological quality.

— The starter culture used should be a commercial lyophilized one. These strains keep relatively well and it is, therefore, advisable to maintain a three-months' supply.

Following the heat-treatment of milk for cheese-making, a certain amount of milk for the preparation of the starter culture is retained in the milk can or in the jacketed vat in order to maintain the heat-treatment of 63°C for an additional 15-30 minutes. The milk is then poured into one-litre bottles and into a 5-litre container and covered with a clean cloth.

Imported Lyophilized Starter Culture

The contents of the vial are poured into a bottle of milk and shaken well to mix the powder and milk together.

Mother Culture

In order to avoid contamination by air, both the bottle containing the mother culture and the 5-litre container of starter culture are placed in a small wooden cupboard.

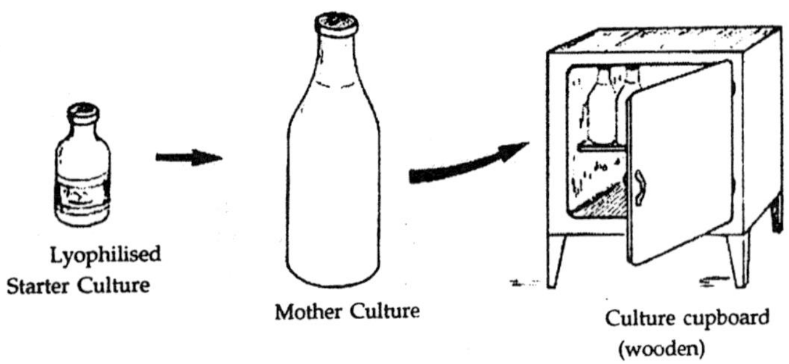

Figure 4.

Starter Culture for Cheese-making

The mother culture to be used for cheese-making (mesophilic culture) is placed in the culture cupboard.

The incubation temperature should be 20 - 22°C for 15-16 hours.

The acidity of the mother culture will then be 80 to 90°D. Use the mother culture to inoculate, at 2 percent, a second mother culture and 5 litres of milk for cheese-making. If the starter cultures are prepared with great care, one strain of commercial culture may be used for one to two months by means of successive subculturings.

Yoghurt-making Cultures

— Starter cultures for yoghurt-making are thermophilic bacteria. They must be cultured therefore at a temperature of 40 - 45°C for three to four hours.

— After heat treatment of the milk to which the yoghurt bacteria are to be added, the milk is cooled to 45°C and then poured into 1 litre bottles. These are then set into a water tank at 45°C to stabilise the bottled milk temperature.

— The lyophilized commercial strain for yoghurt-making is then put into the bottles.

— The bottles of milk are left in the water bath at 45°C for three to four hours.

Preparation of Rennet

Rennet, like starter cultures, may be imported. However, unlike commercial starter strains, rennet can be made locally. Since laboratory equipment is needed for making rennet, it should therefore be undertaken with the advice of a laboratory in the capital city—that for example of a university.

Obtaining Abomasa

Abomasa should be obtained from preferably unweaned calves. Assuming that the annual amount of calf rennet to be obtained is some 120 litres of liquid rennet at a strength of 1/10 000 and a yield of approximately 2 abosama for 1 litre of rennet, 240 abosama will be required per annum.

Preparation of Abosama

- The abosama should be washed and the fat and veins removed.
- The abosama are then inflated with air to avoid the two sides touching. They are then ball-shaped, the neck and the base being tied with string.
- The inflated abosama are hung in a dry, well-ventilated area. Drying should be complete after approximately one month of storage. At this stage, the flattened abosama can be kept in a dry place for a long period without signs of deterioration (about one year).

Soaking

When required for use, the abosamum should be sliced into thin strips, 5 mm wide. In an-easy-to-clean basin of plastic or stainless steel, soak the strips of abosamum in a 10 percent sodium chloride (salt) and 1 percent sodium benzoate solution. To produce a three months' batch of 30 litres of rennet, 60 abosama should be soaked in 40 litres of this brine solution.

The pH of the solution is adjusted to 4.3 with benzoic acid, after soaking for 24 hours at a temperature of from 20 to 25°C. Pour the liquid off into a container. In the same soaking basin and keeping the same abosama strips, the brine solution may be renewed four to five times before the properties of the abosama strips are completely exhausted. Each extract obtained in this fashion is treated as described above and the first and most concentrated extract is used to standardise the strength of the rennet.

Treatment of the Liquid

- To eliminate the mucilage in suspension in the extract, the solution is reacidified with hydrochloric acid to a pH of 4.8, and allowed to settle for two hours.

- The pH of the extract is raised with disodium phosphate to pH 5.5 - 5.6.
- Vigorous stirring must accompany these two actions. The liquid is then filtered over Watmann paper. Filtration may take a long time and probably will require distribution of the extract over several filters.
- The crude rennet extract obtained is generally a golden-yellow colour.

Determination of Strength

Definition: Rennet strength is the number of volumes of coagulated milk clotted by one volume of rennet in 40 minutes at 35°C. If "v" equals one volume of rennet, and "V" one volume of milk and measuring the clotting time in seconds, the calculation is:

$$S = \frac{2400\,V}{Tv}$$

In practice, liquid rennet strength should be 1/10 000 (1 litre of rennet clots 10 000 litres of milk at 35°C in 40 minutes).

Method

Put 500 ml of fresh milk in an Erlenmeyer flask and plunge it in a water bath at 35°C.

Remove 1 ml of the rennet to be standardised and dilute it in 10 ml of water.

When the milk in the Erlenmeyer flask reaches a constant temperature of 35°C, pour 10 ml of diluted rennet, stirring constantly, and start the timer. Keeping the Erlenmeyer flask in the water bath, slant it while rotating it gently so that a film of milk is formed on the sides of the flask. When the liquid begins to floculate, stop the timer.

Rennet strength: $S = \frac{2400 \times 500\ cc}{T \times 1cc}$

If flocculation time is 60 seconds, for example, rennet strength will be:

$$S = \frac{2400 \times 500}{60 \times 1}$$

$$S = 20\ 000$$

Standardisation

The strength of the four or five different batches, obtained by successive extractions of the rennet from the abosama is determined, thus enabling the rennet to be readjusted to a strength of 1/10 000. For example, if 30 litres at a strength of 1/20 000 have been obtained, and the extract from the four preceding rennet mixtures gave a solution at a strength of 1/5 000, the volume of rennet at strength 1/5 000 required to prepare standard rennet at a strength of 1/10 000 is determined in the following way:

30 litres × 20 000 units + "Y" litres × 5 000 units = (30 + Y) litres × 10 000 units

600 000 units + 5 000 Y = 300 000 + 10 000 Y

Y = 60 litres.

After mixing the two extracts, 90 litres of rennet will be obtained at a strength of 1/10 000.

Maintenance-storage

Rennet must be stored in opaque glass bottles or in dark (blue or black-tinted) plastic containers and placed in cold storage at a temperature of 5 to 7°C. Under these storage conditions, the rennet will remain active for three months.

Renneting the Milk

It is recommended that coagulation tests be made first on small volumes of the cheese-milk. This is to redefine the amounts of rennet to use in order to obtain the same coagulation time as obtained previously with powdered rennet of 1/100 000 strength.

TECHNOLOGY FOR THE MANUFACTURE OF DIARY PRODUCTS

Cheese

The following cheese-making operations are for making a firm-bodied cheese (Saint Paulin or Gouda type), a cheese most often produced in the developing countries and one which is relatively easily made to standard. The various steps of the cheese-making process should be adapted to climate, equipment available and consumer preferences.

The producers' milk is first filtered and then poured into the one-step "pasteurization/cheese-making vat". After filtering, a portion of the milk is passed through the separator to remove the required amount of cream. The milk used for cheese-making is standardised to a level of fat content of some 26 g/l. The separator can be a source of contamination, and it is therefore preferable to standardise before pasteurization rather than after.

The milk is pasteurized at low temperature in the double-walled vat at some 63°C for about 30 minutes. The source of heat can be gas or electricity or even fuelwood. The milk is then cooled to a temperature of 32-35°C by circulating cold water through the jacket of the vat. The lactic acid bacteria should be added at a rate of 1 to 2 litres per 100 litres of milk 15 to 20 minutes before renneting. The rennet (at a strength of 1/10 000) is added at a rate of 20 - 25 ml of rennet per 100 litres of milk. This is the time to add, if necessary, calcium chloride (5-50 g per 100 litres of milk).

Following this, flocculation time is from 10 to 15 minutes, and total clotting time is from 15 to 40 minutes. The curd is cut into regular grain-sized pieces. The first stirring should be carried out with both the curd grains and the whey, and should last from 5 to 10 minutes. Drawing off the whey (lactose removal) consists of extraction of a portion of the whey (from 20 to 60 percent), followed by the addition of an equal amount of water at a temperature of 30 to 35°C.

Potassium nitrate may be added at this stage of the cheese-making process. The second stirring, with moderate shaking of the curd grains in the diluted whey, lasts 10 to 20 minutes. The curd and some whey are then put into cloth-covered moulds. These moulds may be made of wood, stainless steel or plastic. Mechanical pressing lasts from 1 to 6 hours. During this operation, the cheese mould will be turned over 2 to 4 times. The cheese is salted in saturated brine at a temperature of 10 to 14°C. Brining time varies according to the volume of cheese.

A Saint-Paulin cheese weighing from 1.5 to 2 kg and with a diameter of 20 cm, is salted for about 8 hours. Working out the curd lasts 2 to 3 days, at a temperature of 10 to 12°C and a humidity rate of 80 to 85 percent. The cheese is placed on wooden shelves to ripen for at least 15 days at a temperature of 10 to 16°C with a humidity rate of 90 to 95 percent. Prior to sale, the cheese can be covered with a protective film of wax. If the cheese processing unit produces very large cheeses, they should be cut into 100 - 200 g slices prior to sale, and wrapped in greaseproof paper.

Cream and Butter

When standardised milk is used to make cheese, there will inevitably be excess fat in the form of cream. Often it is advantageous for the cheese unit to sell this surplus fat as fresh or acidified cream, both of which are more profitable than butter. In rural areas, however, the market for cream is often quite small and the cheese unit is obliged to make butter, a product of longer shelf life.

If the cream is to be sold as it is, it should be pasteurized before packaging. Experience has shown that a pasteurization temperature of approximately 95 - 98°C for 30 seconds destroys germs satisfactorily and inactivates enzymes while preserving the organoleptic qualities of the cream. After pasteurisation, the cream is packaged in plastic bags or tubs and kept in a refrigerator. The fresh cream is usually sold with a fat content of about 40 percent. Acidified cream is sold with a lower fat content - 30 - 35 percent.

To make butter, cream is cooled to the lowest possible temperature. The cream is then stored until enough is obtained to make into butter. Cream stored in this way acidifies automatically, after which it is churned.

After filling the churn, the following steps are taken:

— The churn is rotated at 25 - 35 revolutions per minute (rpm) for 5 minutes.

— The churn is stopped and if necessary the gases released.

— The butter is churned again at 25 - 35 rpm for 35 to 45 minutes.

— The buttermilk is poured off into plastic pails.

— The same amount of cold water as the amount of buttermilk which has been removed is added, and the mixture churned at 10 - 15 rpm for 5 minutes.

— Water is drawn off.

— Rotation at 10 - 15 rpm for 10 to 20 minutes.

— The butter is removed.

— The butter is worked with a butter worker for 5 minutes, compressed into a butter mould, and packaged in greaseproof paper. Alternatively, after working, it is placed in plastic tubs.

The butter can, if required, be salted as it is worked. This operation may be regarded as the standard butter-making process since a number of variations are possible. Non-acidified cream can be churned, in which case the product is called sweetcream butter. Some butter-makers wash the butter twice and in some cases the butter is not worked at all. The larger butter units usually pasteurize the cream, followed by the re-inoculation with selected lactic starter cultures for ripening. This method produces a cultured butter commonly called lactic butter.

Diagram of lactic butter-making:

— Cream

— Pasteurization

— Cooling
— Inoculation with lactic starter culture
— Ripening
— Churning
— Washing
— Working
— Packaging
— Storage and marketing.

Buttermilk

— Buttermilk is a by-product of the butter-making process. It is highly nourishing, hence the interest in promoting the product for human consumption.
— Buttermilk quality depends greatly on the butter-making technique used.
— Buttermilk can be packaged after only one filtration. It should be packaged in plastic bags.
— Buttermilk can also be ripened. While the buttermilk is still in the pail, cheese-making starter cultures (approximately 2 percent) are added, and the product left overnight at room temperature before packaging.

Yoghurt

Steps in yoghurt-making are as follows:

Skimmed or standardised milk

↓

Pasteurisation

↓

Cooling to 42 - 45°C

↓

Inoculation (approximately 1%)

↓

Stirring

↓

Packaging in tubs

↓

Incubation 42 - 45°C for 2 - 5 hours

↓

Cooling

Pasteurisation and cooling are done in the cheese vat, or in a pan for smaller quantities. For yoghurt-making, the dairy unit needs to have a large refrigerator or small cold storage area—this also applies to cream and butter. Another essential investment item is an incubator, and perhaps a hand bottle-capper. Yoghurt is usually sold in plastic tubs or in cartons of 120 or 125 ml.

The yoghurt-making process increases the amount of cream available. Thus the decision to produce yoghurt should be preceded by a cost benefit analysis. Another technique is to let the yoghurt clot in the vat. It is then stirred prior to packaging.

Fermented Milk

The technology is very simple:

Standardisation

↓

Pasteurisation

↓

Cooling to 20 - 25°C

↓

Addition of lactic acid bacteria

↓

Stirring for a few minutes

↓

Ripening (12 - 24 hours)

↓

Packaging

This technology corresponds to the traditional process, and the product is in demand as the base used in preparing "porridge". The processing unit's main consideration in this technique is that the milk is pasteurised and that both quality and hygienic standards are met.

Fermented milk being sold directly at the processing unit does not need to be packaged. The fermented milk is sold from the can, and is ladled into containers brought by local purchasers. No specific equipment is required to make fermented milk. It is, however, advisable to allow for a large refrigerator.

Processed Cheese

Melting emulsification has traditionally facilitated the incorporation of sub-standard cheeses, as well as cheese trimmings, into products for sale as processed cheese. The usual presentation for processed cheese is as a spread, packaged in plastic rolls or aluminium-wrapped in bite-sized portions.

While the latter requires costly equipment to mould and package the cheese, processed cheese spread in plastic rolls offers an attractive approach for small dairies where second-quality cheeses, or cheeses with a manufacturing defect are readily available. The product can be packaged in glass or plastic tubs, or otherwise wrapped in aluminium.

Steps in processed-cheese making

Cheeses

↓

Derinding or scraping

↓

Cutting

↓

Mixing

↓

Addition of emulsifying salt

↓

Cooking (10 min. of which 2 at 90°C; sometimes sterilisation)

↓

Cooling

↓

Packaging when the cheese is still liquid

↓

Final cooling.

Commercial processed cheese preparations are highly diversified, as it is possible to add a variety of ingredients: flavourings, mushrooms, nuts, meats, etc. Packaging while the cheese is liquid, allows great variety of shapes and weights in the finished product. No specific equipment is required to make processed cheese, the only indispensable item being a pot in which to melt the cheese.

Cleaning and Disinfection

Cleaning consists of removing all visible or invisible dirt from the surface. This surface can thus be described as clean. Disinfection involves eliminating or killing micro-organisms. Effective cleaning is absolutely essential for the equipment, installations and premises used for cheese-making, and the cheese-maker needs to pay special attention to this item.

After each use, all equipment and utensils: pails, cans, filters, pans, trays, tables, ladles etc., must be vigorously and meticulously cleaned. This is essential for successful cheese-making as the equipment is the main source of contamination by harmful germs. Simply rinsing with cold or lukewarm water is not sufficient. This is because a very thin film of residues or wastes from the clotting process sticks to the surfaces of the equipment which has come into contact with milk, whey or curd.

To eliminate these residues, an alkaline or acid detergent solution must be added accompanied either by vigorous brushing or agitation of the soaking solution to remove the residual film or wastes. It is recommended that the equipment be soaked immediately after use in a vat filled with water.

Organisation of Work

The layout and design of the dairy unit should be planned in such a way as to allow those responsible for processing operations to work under the confortable, safe and hygienic conditions.

Indicated below, as an example, are the times necessary for each operation involved in processing cheese from 150 litres of milk:

filtering, checking temperature, preparation of coagulation, renneting	60 minutes
moulding, turning	80 minutes
demoulding, salting, setting on trays	60 minutes
Cleaning	50 minutes

To this total of approximately 4 hours, the time required for packaging, labelling and packing the cheese for sale should be added.

References

Levantin, D.L., "Developing beef cattle husbandry", *Zootekhniya,* 1996.

Nasirov, U., "Animal husbandry system, its status and development problems in Uzbekistan", *Dryland Pasture, Forage and Range Network News,* 1996.

Okularczyk, S, Novak, T, Sowula, E., "Cows and beef cattle handling costs with different production scale and mechanization levels", 1996.

Pang, H, Makarechian, M.H, Basarab, J.A., "A simulation study on the effects of cow size and milk production on bioeconomic efficiency of range beef cattle", *Journal of Applied Animal Research,* 1999.

9

Prevention of Cattle Diseases

Cattle diseases are a concern for nearly every beef and dairy producer. Cattle diseases like pneumonia, scours and blackleg can destroy profits practically overnight. Prevention of cattle diseases through vaccination and other management practices can help maximise herd performance. Some of the most common cattle diseases and their control measures and described in this chapter.

Abortion

Cows can suffer abnormalities during pregnancy leading to mummification of the foetus or resulting from maternal or foetal abnormality. All cases where the pregnancy terminates early and the foetus is expulsed are called abortions.

As there are multiple causes of abortion and the detection of abortions in a herd can vary significantly depending on the husbandry system and calving pattern, the incidence of abortion at herd level also varies markedly. It has been suggested that an abortion rate of 5% or more in a herd should be considered an indication of an abortion problem.

The causes of abortion have been classified by Boyd and Gray as follows:

- Infectious:
  - Non-specific
  - Specific

— Miscellaneous:
  — Drug-induced (prostaglandins)
  — Insemination/intra-uterine infusion
  — Hypothyroidism
  — Trauma/stress (transport, noise, veterinary treatment etc.)
  — High fever and endotoxins (toxic plants, nitrate/nitrite, fungal toxins, other disease)
  — Nutritional (malnutrition, vitamin A/selenium/vitamin E deficiency, goitre)
  — Twin pregnancy
  — Genetic (malformation)

Both non-specific and specific infectious causes of abortion can lead to "abortion storms" in a herd, whereas the miscellaneous causes often result in sporadic, individual cases.

The most important infectious abortion agents are:

— Salmonellosis (particularly Salmonella dublin)
— Listeriosis
— Leptospirosis
— Neospora caninum abortion
— Bovine Viral Diarrhoea (BVD)
— Infectious Bovine Rhinotracheitis (IBR)
— Campylobacteriosis
— Fungal/mycotic abortion
— Epizootic/chlamydial abortion
— Trichomoniasis
— Brucellosis

Campylobacter jejuni is seldom reported as a primary cause of abortion in cattle in the UK. This may be due to the limitations of current diagnostic procedures. It has been suggested that the pathogen may be a significant cause of infertility in suckler herds.

Fungal abortions are associated with contaminated feed and are either sporadic or occur in small outbreaks, typically between the fifth and seventh months of pregnancy.

Epizootic abortion in cattle, caused by Chlamydia spp., has been reported in the UK in recent years and is seen as an emerging cause of contagious abortion in cattle.

Abortions caused by trichomoniasis and brucellosis are extremely rare in the UK, as both diseases have been subject to statutory disease control measures. Brucellosis is a notifiable disease and its control is still carried our as a statutory measure under the Brucellosis Order.

Control and Prevention

Due to the multiple aetiology of abortions in cattle, general prevention of non-infectious abortion concentrates on good husbandry, feeding and management. The control of infectious abortion is based on good disease control through closed herd policy, careful screening and quarantine of bought-in or introduced animals and good biosecurity. Specific control measures for individual abortion agents are presented under each disease.

Once a sporadic abortion or an outbreak of abortions has happened, it is, in most cases, difficult to prevent further abortions from occurring unless the causative agent is identified and can be eradicated. In some cases, like IBR, further abortions will occur months after the first event. An important cornerstone of abortion control in a cattle herd is good record-keeping of abortion events, and identification, if possible, of the causes in each detected case of abortion. It is, therefore, good practice to investigate all abortions, even though only about one third of all laboratory investigations of abortions produce a conclusive result.

Over a period of time, a picture of the herd's status will emerge, and subsequent abortions may be diagnosed tentatively based on clinical signs only. A systemic examination of abortion cases should be carried out by a veterinary surgeon who will collect information on the history of the individual cow and the herd, examine the cow and the foetus (including placenta), collect laboratory samples and interpret results in connection with the cow and herd history.

Methods of Treatment

As it is virtually impossible to prevent an abortion from happening once the expulsion of a foetus has began, the treatment should concentrate on making sure that the dam does not suffer any long-term effects from the event. Subsequent retained foetal membranes or metritis need to be attended to. In the case of an unknown cause, further spread of a potential contagious infection should be prevented by isolating the affected cow/cows.

To prevent abortions from happening or abortive agents from entering the herd:

- Establish a closed herd policy.
- Screen and isolate bought-in animals.
- Establish good biosecurity on the farm.
- Establish a good recording system for abortions on the farm (date, cow, stage of gestation, additional information, results of systemic examination by a vet).
- Get a veterinarian to carry out a systemic examination of all cases of abortion in the herd if the annual incidence is more than 4% (4 cows in 100).

Anthrax

Anthrax is an acute disease in humans and animals that is caused by the bacterium Bacillus anthracis and is highly lethal in some forms. Anthrax is one of only a few bacteria that can form long-lived spores. When the bacteria's life cycle is threatened by factors such as lack of food caused by their host dying or by a change of temperature, the bacteria turn themselves into more or less dormant spores to wait for another host to continue their life cycle.

Clinical Signs

- Sudden death (often within 2 or 3 hours of being apparently normal) is by far the most common sign

— Very occasionally some animals may show trembling, a high temperature, difficulty breathing, collapse and convulsions before death. This usually occurs over a period of 24 hours
— After death blood may not clot, resulting in a small amount of bloody discharge from the nose, mouth and other openings

Diagnosis

— Rod-shaped bacteria surrounded by a capsule are visible in blood smears made from surface blood vessels.
— Post-mortem examinations should not be undertaken on suspected anthrax cases (including any cow that has died suddenly for no apparent reason) until a blood smear has proved negative).
— If a carcass is opened accidentally, the spleen is usually swollen and there is bloodstained fluid in all body cavities Suspected anthrax cases are covered by the Anthrax Order 1991 (which replaced the Order of 1938). Any suspicion of the disease must be notified to the DEFRA Divisional Veterinary Manager (DVM), who if they see fit will instigate a veterinary enquiry to determine whether anthrax is present on the farm. Usually the DVM will arrange a visit to take a blood sample to look for bacteria with capsules. The animal or carcass must not be moved before this has been done, fines will be levied if movement does occur. If the tests prove negative, the veterinarian will send in a certificate stating this and nothing further will happen. If anthrax is still suspected then orders banning movement and requiring disinfection will be instigated.

Treatment

— Due to the rapidity of the disease treatment is seldom possible
— High doses of penicillin have been effective in the later stages of some outbreaks

Prevention

Infection is usually acquired through the ingestion of contaminated soil, fodder or compound feed. Sterilisation of meat and bone meal used in animal feed (and more recently its complete removal from cattle feed) has been the main factor responsible for the rarity of anthrax in the UK over the last few years.

Lice in Cattle

The NADIS data show that last winter there was an increase in the number of cases of lice seen by NADIS vets. Lice populations are highest in winter and lowest in summer. There are two main reasons for this. Firstly, housing significantly increases the rate of transfer of lice between cattle. Secondly, low light levels and cooler skin temperatures are associated with increased louse activity. The denser winter coat and cooler weather thus favours lice survival.

Effect of Lice

Lice cause irritation of the skin. This leads to biting, scratching and rubbing by affected cattle. These cows may also damage fences, trees and buildings while rubbing. The effect of lice on the production and growth rate of cattle has been the subject of much research but is still is a matter for continued debate. Their effect on the skin of cattle is probably best understood.

Lice are probably the primary cause of "light spot and fleck", a blemish visible on the hide of cattle which down grades the value of leather and is estimated to cost the leather industry £20 million per year. Other effects such as weight loss, poor milk production and anaemia are less proven, even when there are large numbers of lice involved. This is probably because large numbers of lice are usually seen in animals that are under stress or under-fed or that have other current disease, which can all result in poor productivity without lice involvement.

Types of Lice

There are four common species of lice in the UK, which can be divided into two different categories:

— *Sucking Lice*: There are 3 species commonly found in the UK. These have relatively small narrow heads designed piercing the skin and sucking blood. In large numbers they can cause anaemia. They are usually found around the head and neck of cattle.

— *Biting Lice*: Biting lice have larger rounder heads. They feed on skin debris, blood and scabs. Despite being apparently less invasive than sucking lice, it is biting lice that produce the most severe irritation. There is one species of biting louse found throughout the world. It is a reddish-brown louse about 2 mm long with a brown head. It is mostly found on the neck, shoulders, back and rump.

The life cycles of all species are similar. The female lays a few hundred eggs over the period of one month. These eggs are glued to the hair shafts, and hatch within a few days as nymphs (which resemble small soft adults). These develop, grow and moult three times before they become adult, with each stage lasting approximately one week. The entire life cycle takes between three and six weeks.

Diagnosis

On clinical signs and finding one of the three stages of the life cycle. Eggs are usually the easiest stage to spot, being found on hairs adjacent to bald, rubbed areas. Careful examination of nearby skin, with a magnifying glass, will usually detect nymphs and adults.

Treatment

Lice are spread only by direct contact between cattle. Adults, nymphs and eggs cannot survive more than a few days if removed from cattle. If properly applied treatment can eradicate lice from a farm. Most insecticides are effective against adult lice and nymphs. However most are not very active against louse eggs. This means that after treatment, eggs can still hatch and continue the infestation, unless there is some residual action. Ask your vet for advice as to which product has the best persistence.

It can be important to know whether you have sucking or biting lice, because the different method of feeding means that they have different susceptibilities to treatments. This is particularly important if you are going to use an avermectin injection as these are much more effective against sucking than biting lice. If you want to use such a product ensure you have the lice on your cattle identified. The timing and frequency of treatments depend very much on individual circumstances. In many cases treatment in late autumn or early winter will give adequate control of cattle lice for the whole housing period.

Whichever product you use, dose accurately, ensuring that you do not under-dose as under-dosing is the best way of ensuring the development of lice that are resistant to treatment. Treat all cattle on the property at the same time if possible, choosing a time when they are not stressed or in poor condition. If groups have to be treated separately, such groups should be kept apart to ensure there is no contact between treated and untreated groups.

For cattle that have light to moderate numbers of lice, treatment cannot be justified in terms of improving growth rate, body condition or productivity. Treatment can improve hide quality, but as yet this is not of economic importance in the UK. However some quality assurance programmes in other countries have made hide quality of economic importance and it is possible that this will come in the UK. Treatment may also become necessary on welfare grounds because of the easily appreciated discomfort that even moderate lice infestations cause.

Listeriosis

Listeriosis is an infectious disease caused by a bacterium, Listeria monocytogenes. This is a zoonotic disease that can be spread from animals to humans and has been implicated as a potential human health risk, associated with consumption of contaminated milk or meat.

In cattle, the most common manifestations of listeriosis are meningoencephalitis, abortion, mastitis and septicaemia. Very little current information on the prevalence of Listeria infections in British cattle is available. Earlier surveys from the 1970s and 1980s suggest that abortion is the most common form of disease in cattle, with encephalitis and mastitis occurring more sporadically. Skin infection with L. monocytogenes, cutaneous listeriosis, has also been reported in the UK. In recent years, an increasing number of cases of uveitis, iritis and keratoconjunctivitis caused by L. monocytogenes of silage origin have been reported.

Many animals excrete L. monocytogenes in their faeces as a normal part of the intestinal flora. Oral infection via contaminated silage is the most common route of infection, but nasal or venereal infections can also occur. This type of infection is seen most often in 1-2-year old animals as result of cutting of molar teeth. There are a number of predisposing factors that cause disease by the agent. Nutritional deficiency, particularly malnutrition, poor quality silage (high pH), heavy silage feeding, cold and wet periods, stress caused by long transport, potential feed contamination by vermin and the entry of clinically normal carrier animals into a clean herd have been recognised as risk factors for listeriosis.

The bacteria persist up to two years in slurry, manure and straw and appear to persist better in cold than warm conditions. Abortion caused by L. monocytogenes tends to occur in late gestation (6-8 months), during the winter months, is sporadic by nature and recurs year after year.

Certain serotypes of L. monocytogenes are particularly associated with listerial encephalitis. Many reported cases are associated with fermented silage and are sporadic. Listerial encephalitis symptoms can be confused with those of BSE. Chronic mastitis caused by L. monocytogenes has also been reported in dairy cows and is of particular concern for farms that process unpasteurised milk for consumption.

Control and Prevention

As most cases of listeriosis are associated with feeding soil-contaminated, fermented or poor quality silage with high pH-values, feeding good quality silage is the most commonly recommended method of prevention in regard to listeriosis. Making sure that silage is not contaminated with excessive soil during silage making and removing contaminated or fermented parts of the silage as they are detected are likely to reduce the risk for listerial infections.

Feeding fattening stock with in-feed antibiotics on a prophylactic basis is not acceptable under organic standards. Once an outbreak or a sporadic case has occurred and been diagnosed, the veterinarian might want to consider metaphylactic treatment of in-contact animals with suitable antibiotics. This is acceptable under organic standards as the welfare implications of further cases would be serious.

Treatment

The treatment of listerial encephalitis and mastitis has been attempted with antibiotics. Most antibiotics are not effective against L. monocytogenes, however, and veterinary advice should be sought. The prognosis is poor in most cases and destruction of the infected animals should be considered. Listerial abortions often lead to other reproductive problems and culling of the animal is usually advisable. Steroids or atropine are common treatments for listerial eye-infection.

Mange in Cattle

Mange is the term used to describe infection by mites, microscopic relatives of spiders. They inhabit and damage the skin of domestic animals and man. Problems are most frequently seen in the autumn and winter but can occur all year round. There are three main species of mite that affect cattle in the UK, the surface mite (Chorioptes bovis), the burrowing mite (Sarcoptes scabiei) and the sheep scab mite (Psoroptes ovis) The surface mite is the most commonly seen in the UK.

Effect of Mites

The surface mite is usually found on the neck, legs, and tail head. It produces limited hair loss, which only increases slowly in size. However, the lesions are obviously itchy which results in hide damage elsewhere as the cattle try to rub the affected areas. The sheep scab mite is found on the flanks and around the tail head and anus. Although this mite feeds on the surface of the skin, its mouthparts pierce the skin, producing blisters, which are very irritant.

The burrowing mite prefers the neck and the loin area next to the tail (leading to the description of 'neck and tail' mange). As they burrow into and out of the skin they produce a much more intense irritant reaction so that the skin damage rapidly develops with much larger areas being affected and the skin becoming very thickened and crusty. Infection of the damaged areas often develops and affected animals have much reduced production.

Life Cycle

The surface mite and the sheep scab mite both spend their entire life cycle on the surface of the skin. Females lay around 90 eggs which once hatched take around ten days to develop into mature adults. The burrowing mites lifecycle is more complex. The female mite tunnels into the skin, and lays around 50 eggs. These hatch in four or five days, each releasing a larva. Some of these tunnel to the surface to become adult others develop in the tunnels; this process takes around two weeks. More tunnels are often formed during the mating process.

For all three species, infection is spread mainly by direct contact between cattle. However, the burrowing mite can survive for some time off the host, so, for this species, bedding and objects that come into contact with infected animals may become contaminated and help spread the infection. For the sheep scab mite, although mites found on cattle are very similar to those found on sheep, it is very unlikely that natural spread from cow to sheep occurs.

Diagnosis

Areas of thickened skin in obviously itchy animals are very suggestive of mites, particularly if there is no evidence of lice. To confirm a diagnosis get your vet to take a skin scraping for examination under a microscope.

Treatment

A range of products is available to treat mange in cattle. The choice is between pour-on products and injections. The first are easier and quicker to use and are often cheaper. However, in severely infected animals, the skin reaction can mean that contact between the product and the mite is limited. In such cases, scabs may have to be removed before treatment. If very severe then injectable products are probably a better bet.

For very severe surface mite problems, an injection should be followed up by a pour-on treatment when the skin has recovered, as in this species injections only control but do not eliminate. Sheep scab mite can be effectively controlled with injections. The timing and frequency of treatments depend very much on individual circumstances. In most clinical cases, two treatments will give adequate control of cattle mites for the housing period.

Whichever product you use, dose accurately, ensuring that you do not under-dose as under-dosing is the best way of ensuring the development of mites that are resistant to treatment. Treat all cattle on the property at the same time if possible, choosing a time when they are not stressed or in poor condition. If groups have to be treated separately, such groups should be kept apart to ensure there is no contact between treated and untreated groups.

Mastitis

Mastitis is one of the three most significant health problems of the UK dairy herds, together with lameness and fertility problems. The effect of mastitis on milk quality has recently gained added importance with the introduction of an EU-wide upper legal limit for somatic cell counts in milk destined for liquid market at 400 000 cells/ml. Public interest in the welfare of production animals and the recognition of mastitis as a major source of pain for affected cows give added focus to mastitis concerns.

Organic dairy farmers have identified mastitis as a major concern in the UK, mainly due to non-use of antibiotic dry cow therapy and the need to maintain low somatic cell counts in the milk. Animal welfare aspects of controlling and treating mastitis are also important on organic farms, where the maintenance of high welfare standards is important.

Causative Factors

Our understanding of mastitis has developed in several stages over the past 100 years. An association between mastitis and pathogenic micro-organisms was established in 1887. Most major pathogens were identified by the 1940s. When antimicrobial therapy became available for production animals in 1945 it proved effective in the control of some, but not all, mastitis pathogens. This prompted further research into potential husbandry related causes of mastitis.

In the 1960s, the multifactorial aetiology of bovine mastitis was commonly recognised. Today, mastitis is considered to be a multifactorial disease, closely related to the production system and environment that the cows are kept in. Mastitis risk factors or disease determinants can be classified into three groups: host, pathogen and environmental determinants.

Somatic Cell Counts (SCC)

Somatic cell counts (SCC) have long been used as a way of measuring milk quality. Most dairy companies base their milk pricing policy, among other things, on SCC values of the milk. The SCC levels in the national dairy herd in the UK have declined steadily since the 1970s and are now well below 200 000 cells/ml, both in bulk tank milk and in average individual cow milk in milk-recorded herds. The maximum legal limit for saleable milk is 400 000 cells/ml.

The somatic cells consist mainly of immune cells that enter the milk compartment of the udder. Only a minority of these cells are dead cells from the udder tissue. There are always small quantities of immune cells in the cow's milk, and their function is to protect the udder against infection by bacteria. The older the animal gets, the more somatic cells it tends to have in its milk. Similarly, the SCC levels are higher immediately after calving and towards the end of each lactation. When bacteria do enter the udder, the number of immune cells increases rapidly, as the immune system attempts to overcome the infection.

Once the infection has been cleared, the SCC levels gradually drop to normal. This can sometimes take weeks, however. In cases of chronic infection, where the bacteria persist in the udder, the SCC levels can remain high throughout the lactation. High SCC levels in the milk cause deterioration of the milk quality. It has been shown that levels above 500 000 cells/ml decrease cheese yield and affect yoghurt making.

The shelf life of milk is also affected, but at a higher level of SCC. Consistently high SCC levels in a herd are usually a sign of high levels of subclinical mastitis. Most cases of subclinical mastitis are caused by contagious mastitis bacteria, even though Str. uberis is increasingly considered to cause chronic mastitis as well.

Control and Prevention

Mastitis is a multifactorial disease, closely related to the production system and environment in which the cows are kept. Mastitis risk factors or disease determinants can be classified into three groups: host, pathogen and environmental determinants.

Host Determinants

Genetic Resistance

Genetic resistance to mastitis has been well researched. There is a body of work on the genetics of SCC and subclinical and clinical mastitis. This work has established a favourable genetic correlation between low SCC and mastitis incidence at the cow level. Others have reported on the significance of certain BoLA alleles in resistance to Staphylococcus aureus-infection. A particular concern is that factors increasing resistance to one udder pathogen might predispose to mastitis caused by other pathogens.

In the past, efforts to introduce mastitis resistance traits into dairy cow breeding schemes have also been hampered by the negative genetic correlation with increased milk yield. It has often been considered uneconomic to attempt to improve both mastitis resistance and milk yield simultaneously, particularly since the heritability of milk yield is markedly higher than that of mastitis resistance. Recent introduction of bull proofs for SCC by the Animal Data Centre in the UK allows farmers to choose bulls for AI that will produce daughters with lower SCC profiles.

SCC in the Milk

The correlation between SCC in the milk and the immune response of the udder to infection is not clear. There is evidence to suggest that minor pathogens increase milk cell counts and can help to protect the udder against mastitis. For example, it has been shown that Corynebacterium bovis-infection protects the udder against infection by major pathogens.

Several field studies have also concluded that low BTMSCC herds have a higher incidence of environmental mastitis compared to herds with high BTMSCC. There is also experimental evidence to suggest that moderate to high individual cow milk SCC can provide protection against experimental infection by environmental mastitis pathogens.

Cow and Udder Conformation

Udder and foot conformation have been shown to be important risk factors for mastitis. Most conformation related traits have high heritability, and are generally recognised by farmers and herdsmen as major selection criteria for breeding. There is an emphasis on appropriate breeding in organic standards, and hence breed may be an important factor influencing both mastitis resistance and susceptibility characteristics.

Milk Yield

There is substantial evidence to suggest that high yields are linked to high mastitis levels.

Nutritional Status

Various nutritional factors may lower a cow's disease resistance. The phagocytic activity of macrophages in milk is significantly inhibited by ketone bodies. Several studies have shown that the outcome of experimental E. coli-infection is related to the in vitro ability of polymorphonuclear cells to react to chemo-attractants. Energy, protein and mineral/trace element deficiencies may also affect disease resistance and SCC levels. Baars & Opdam found a reduction in CMSCC in heifers when the animals received a bolus of trace elements and vitamins. Selenium supplementation has been linked to improved immunity to mastitis.

Disease surveys of organic dairy herds both in the UK and in Denmark and Norway suggest that ketosis occurs at the same or a lower level than data from conventional farms indicate. Organic nutritional management of dairy cows may increase the risk of mastitis by causing unrecognised mineral deficiencies in the absence of routine supplementation. On the other hand, organic feeding restrictions are less likely to lead to high yields and ketosis, decreasing mastitis risk.

Age of Host

Both mastitis incidence and SCC levels are higher in older cows. Organic livestock production standards do not directly discourage culling of livestock for age or parity, but positive animal health and welfare aim at longevity, and the culling of healthy young animals would probably be considered as contradictory to positive welfare management. There is very little information on the longevity of organic dairy cows. A UK survey of 16 established organic farms found an average parity of 3.1 on these farms.

Stage of Lactation

Most mastitis surveys show that 2/3 of all clinical mastitis cases occur in early lactation. Mastitis research carried out on Danish organic farms did not find any increase in dry period mastitis, whereas a UK survey found that 50% of the surveyed farms had relatively high levels of dry period mastitis in comparison with conventional farms that had virtually no dry period mastitis. Organic standards prevent the use of routine antibiotic DCT on dairy cows. It is possible that British dairy farmers, who have been accustomed to use antibiotic DCT on all cows at drying-off, find it more difficult to dry cows off without this prophylactic support than their Danish counterparts who have never used DCT on a routine basis.

Other Diseases

It is recognised that other diseases, particularly ketosis, milk fever, lameness and post-puerperal endometritis/metritis, are closely associated with mastitis incidence. Information from surveys of disease incidence or prevalence on organic dairy farms from the UK and other European countries suggests that disease levels are either similar or, particularly in the case of ketosis, lower than on conventional dairy farms.

Vaccination Status of Host

E. coli is the only mastitis pathogen that has been commonly vaccinated against in the US. Vaccines based on R-mutant coliforms have been shown to lower the severity of clinical symptoms but appear to have no effect on the prevalence of infection. An *E. coli* vaccine has recently been introduced to the UK market. Routine use of such a vaccine, particularly since there is some doubt as to its efficacy in preventing infection, would not be acceptable in an organic herd. It could, however, be used as a part of a disease reduction strategy alongside other management and husbandry improvements in herds where E. coli has been identified as a major problem.

Pathogen Determinants

Mastitis Pathogens

Whilst over 100 different micro-organisms have been identified as causative agents of mastitis, only a few species of staphylococci, streptococci and Gram-negative organisms are of economic or epidemiological significance. The importance of the various mastitis pathogens has also markedly changed throughout the past 50 years as a result of different control and husbandry methods used. Major mastitis pathogens are classified as being either environmental or contagious.

The routine use of antibiotics and improved understanding of the complex aetiology of mastitis have meant that the targeting of control, and even eradication, of some mastitis pathogens, has became more efficient. Increased emphasis on somatic cell count reduction and targeting certain contagious micro-organisms may have changed the relative importance of the principal mastitis pathogens in the national herd.

Low SCC herds may be more susceptible to environmental mastitis caused by Escherichia coli, which are becoming more important, whilst Str. agalactiae is rapidly disappearing. This phenomenon is coupled with apparent changes in the virulence of some pathogens (*Str. uberis*) and with the emergence of previously non-pathogenic or minor pathogens (coagulase-negative staphylococci) as mastitis causing pathogens. There is also increasing evidence that bacteria that until recently have been considered non-pathogenic or opportunistic udder pathogens are becoming more common as primary mastitis pathogens. These bacteria include Corynebacterium bovis and coagulase negative staphylococci.

Whilst mastitis levels in organic herds have been studied in various European countries, including the UK, very little information exists on the occurrence of different mastitis pathogens in organic dairy herds. Hovi and Roderick found no difference in the prevalence of different pathogens in a limited sample of organic and conventional farms. It has been suggested, however, that udder infections with contagious pathogens, particularly those that show fewer clinical signs, would become more important in organic herds where blanket antibioitc dry cow therapy is not used.

Contagious Mastitis Pathogens

The contagious pathogens usually have a mechanism to adhere to the epithelial cells of the udder or to become intracellular, in order to protect themselves from the intramammry defense mechanisms. *Staphylococcus aureus, Streptococcus agalactiae* and *Streptococcus dysgalactiae* belong to this group of pathogens. *Actinomyces pyogenes* is an opportunistic, contagious mastitis pathogen, usually spread by flies.

Mastitis caused by these microbes is often chronic and causes elevated SCC levels. It is possible to eradicate contagious mastitis pathogens from a herd by aggressive antimicrobial therapy and/or culling and biosecurity. Antibiotic dry cow therapy has been seen as a major factor in diminishing the significance of these pathogens in British dairy herds.

Environmental Mastitis Pathogens

Environmental mastitis bacteria include a large number of both Gram-positive and Gram-negative species. Str. uberis, Str, equinus, *Enterococcus faecalis* and *Enterococcus faecium* of the Gram-positive species and *Escherichia coli, Klebsiella* spp., *Enterobacter* spp., *Serratia* spp. and *Pseudomonas* spp. of the Gram-negative are the most common environmental pathogens of the bovine udder. *Str. uberis* and *E. coli,* however represent by far the largest proportion of the identified intramammary infections caused by environmental pathogens in the UK.

The significance of *Str. uberis* and *E. coli* has grown in the past 15 years as *Str. agalactiae* and *S. aureus* have been controlled successfully in many herds. It has also been suggested by various surveys that these pathogens, particularly *E. coli*, have become more significant in herds with low somatic cell counts. It is estimated that currently approximately 20% of mastitis cases in the UK are caused by Str. uberis, and a similar proportion by *E. coli*.

Whilst it has been recognised that teat injuries, wet bedding and contamination with faecal material are important risk factors for *E.coli* mastitis, it has also been suggested that improvements in hygiene are not reducing *E. coli* mastitis incidence, as cows are becoming more susceptible to the disease. One of the reasons for the increased susceptibility is likely to be the increased milk flow capacity in cows, leading to milk leaking.

Opportunistic Udder Pathogens

Opinions on the significance of Corynebacterium bovis as an udder pathogen vary greatly in the literature. Some workers suggest that, whilst intramammary infections with *C. bovis* cause increased SCC in affected quarters, the presence of this minor pathogen provides protection against major pathogens. Others, however, find no protective effect.

In many countries with intensive dairy production, coagulase negative staphylococci (CNS) have been identified as emerging mastitis pathogens, suggesting that increasing numbers of bacteria considered non-pathogenic, until recently, are capable of causing clinical itramammary infections. Generally, it is accepted that the mastitis caused by these organisms is mild or subclinical. It has been suggested that the susceptibility of dairy cows to mastitis caused by CNS is a reflection of lowered resistance in the cow's udder.

Environmental Determinants

Organic standards recommend the use of straw or other appropriate bedding material and require that all animals have access to dry lying areas. Loose housing is also a requirement. Whilst a requirement for drying lying areas is likely to decrease the risk of mastitis, loose housing on straw yards is likely to increase the risk of environmental mastitis.

The most likely risk factor for introducing new mastitis pathogens into a dairy herd is a new cow or heifer that carries an infection. A closed herd policy is the best safeguard against this risk and organic standards recommend limiting the number of animals brought in annually to 10% of the herd.

Culling Policies

The presence of chronically infected cows in a dairy herd is a well-recognised risk factor for mastitis. Due to poor cure rates with antibiotic treatment and due to their contagious nature, *S. aureus* infections are seen as particularly dangerous.

Mastitis Treatment Practices

Mastitis treatment practices can affect the transmission of pathogens within the herd. If the main aim of the treatment is not to eliminate the pathogen as quickly as possible, the duration of infection increases, increasing the risk of transmission. A UK survey of organic dairy farms revealed no difference in mastitis levels between the farms that used primarily antibiotics and those using primarily alternative therapy, mainly homeopathy.

The fact that antibiotic DCT is not used as a method of treatment for chronic cases of mastitis or for cows that have an udder infection at drying-off is likely to increase the duration and prevalence of udder infections in the herd. This may be reflected in higher subclinical mastitis levels and a higher dry period incidence in organic herds.

Milking Hygiene

The main aim of milking hygiene is to prevent the spread of contagious mastitis from one cow to another and the introduction of environmental or contagious bacteria inside the teat canal during milking. The most effective way of avoiding these risks is to milk infected cows separately or last and to keep udders, teat and the milking machine clean. None of the UK organic dairy herds surveyed by Hovi and Roderick practised separation of infected animals at milking.

Disinfecting of Teat

Teat disinfecting or teat dipping after milking has a major effect on the microbes growing on the teat. The dipping practice was first introduced to prevent the spread of contagious pathogens during milking. Disinfecting of teats before milking has also been recommended as a way of preventing the spread of contagious pathogens during milking.

Whilst these methods have been very successful in combating contagious mastitis, it has been suggested that, as well as killing off the pathogenic microbes, the disinfectants also destroy other microbial flora that function as "healthy" competition against colonisation of the teat, particularly by environmental mastitis-causing organisms.

The organic standards discourage the use of chemicals and encourage natural resistance to disease. However, most organic dairy farmers surveyed by Hovi and Roderick used post-milking teat dipping, at least during the housing period, and some even used pre-milking teat disinfection.

Teat Injuries

Teat injuries are also likely to lead to improved survival of pathogens on the teat. The main causes of teat injuries tend to be lameness and inappropriate housing systems. Existing information from organic dairy herds in the UK and elsewhere suggests that lameness is not a specific problem for organic dairy herds.

Udder Cleanliness

Udder cleanliness is an important factor in the general resistance to mastitis. Dirt on udders and teats increases infection pressure, damages skin and prevents beneficial, commensal flora from establishing. As organic standards require provision of dry bedding areas, udder cleanliness is enhanced. In a UK survey of organic dairy farms, housing conditions were scored above average in most herds.

Milking Machine and Milking Technique

Milking machine faults and poor milking techniques are probably among the main environmental risk factors for mastitis, alongside housing hygiene. Unstable or excessive vacuum, faulty pulsation, liner slippage for various reasons and teat cup hygiene contribute to mastitis risk by either damaging the patency of the teat canal or by causing pathogenic organisms to enter the teat canal during milking. The organic production standards do not have any special implications for milking techniques or milking machine maintenance. In a UK survey of organic dairy farms, all farms had their milking machines tested twice per year or more often.

Husbandry Practices

The teat canal remains open for up to 45 minutes after milking. A recommended husbandry practice is to prevent the cows from lying down until the teat canal has closed, to prevent bacterial penetration. In a UK survey of organic dairy herds, only 6 out of 16 organic farms prevented cows from lying down directly after milking.

Stockmanship

Stockmanship and other characteristics of the herdsman have been considered important enough by some workers to warrant a separate category among disease determinants. There is, however, very little published information on mastitis and stockmanship. Organic standards offer little guidance on stockmanship. The human-animal relationship is mentioned in the old UKROFS standards, specifying that animals be housed in conditions allowing them regular sight, smell and sound of human activity, but this has been removed from the current standard.

Conventional Approach to Mastitis Control and Prevention

For the past 25 years, the theory and practice of conventional mastitis control in the UK have been based on the Five Point Plan, developed at the National Institute for Research in

Dairying in conjunction with the Central Veterinary Laboratory, in the 1960s.

The Five Point Plan For Control of Mastitis in Dairy Herds:

- Routine post-milking teat dipping.
- Prompt treatment of clinical mastitis with antibiotics.
- Blanket antibiotic dry cow therapy for the whole herd.
- Culling of cows with chronic mastitis.
- Milking machine maintenance with annual testing.

The Five Point Plan is unsuitable for organic dairy farms, due to its emphasis on antibiotic DCT for every cow at drying-off as well as on the elimination of existing infections by antimicrobial therapy and culling. Strict control of somatic cell counts by antimicrobial prophylaxis and aggressive culling does not fit with the organic principles of positive health care, reduction of chemical inputs and animal welfare. Furthermore, the Five Point Plan offers very little guidance on disease prevention strategies recommended by organic standards or on mastitis treatment without antimicrobials.

Fatty Liver Syndrome

Fatty liver syndrome is the accumulation of fat within the cow's liver. The dairy cow does not normally store fat in the liver, so fatty liver does not occur when a cow increases its body condition and puts fat on its back. Fatty liver occurs as a result of the cow breaking down too much fat for the liver to process properly, the broken down fat products are then converted back to fat in the liver to prevent them becoming toxic. Thus the liver becomes fat when the cow is losing condition, the more loss in condition the more fat in the liver.

Fatty liver syndrome reduces liver function, depresses appetite and milk yield, increases the risk of diseases such as RFM, metritis and mastitis, reduces fertility, and when severe (when it is usually called fat cow syndrome) can lead to death. Once it is deposited in the liver, the concentration of fat in the liver does not fall until the cow gets into positive energy balance, which can be over ten weeks after calving, particularly if the fatty liver is severe. Fat cows are much more prone to fatty liver.

Causes of Fatty Liver

Fatty liver, ketosis and displaced abomasum are closely interconnected. Cows which have one of these conditions are much more likely to get another. The most important cause of fatty liver is negative energy balance.

Clinical Signs

- High incidence of diseases such as milk fever, ketosis, mastitis after calving.
- Reduced fertility.
- Rapid weight loss after calving particularly in cows that were fat at calving.
- Reduced milk yield (often on a herd basis).

Diagnosis

- Clinical signs
- Blood samples: increased NEFA (free fatty acids), increased ketones (such as beta-hydroxy butyrate), increased liver enzymes Liver biopsy: this the best diagnostic test.

Treatment

- Use the same treatment as for ketosis: Glucose, propylene glycol, corticosteroids.
- However treatment is often ineffective.
- Prevention is far more important.

Prevention and Control

Fatty liver occurs because of too much fat breakdown after calving. This occurs primarily in cows that are too fat at calving. Therefore ensuring that cows calve at a body condition score between 2.5 to 3.0 will significantly reduce the risk of fatty liver. Cows should be dried off at a body condition score of 2.5 to 3.0 and maintain their body condition during the dry period. Any alteration of body condition score is best done during mid to late lactation.

Downer Cow

The downer cow is not very well defined. It is simply a cow that on examination ought to rise but doesn't. There are many causes of a downer cow, including:

- *Trauma at or after calving*: Bone fracture or nerve paralysis
- *Metabolic*: Milk fever or hypomag
- *Toxic disease*: Metritis or mastitis

A cow becomes a downer cow when the initial cause resolves but the cow still doesn't rise. This failure to rise is usually observed within 24 hours of the cow going off her feet, as a result of muscle and nerve damage. This damage occurs because a cow going off its feet results in heavy pressure on its muscles and nerves, this is made worse in many diseases by the cow being unable to shift position to prevent continuous bearing of weight.

Clinical Signs

- Recently calved cow (usually less than 48 hours).
- Unable to rise for no apparent reason.
- Lie in sternal recumbency (on the breast bone).
- Alert, will often eat and drink and pass urine and faeces.
- Most make no effort to rise, but some move around on forelimbs (creeper cows).

Diagnosis

- On the clinical signs described above.
- The downer cow is a diagnosis of exclusion, so a veterinary examination is essential to rule out broken bones, nerve paralysis, unusual milk fevers, metritis etc.

— Blood tests can be very useful in assessing the prognosis, as can the presence of reflexes.

Treatment

— Move to a well bedded yard or loose-box if housed.

— Good nursing care is the key to success, e.g. providing food and water in easy to reach wide-based containers.

— Mechanically raising the cow, followed by hobbling, can help in many cases.

— Observe closely for toxic mastitis, as this is very common, even in cows which did not have mastitis originally.

— Give calcium, phosphorus and magnesium as necessary.

— Local disinfection and treatment are necessary in more severe cases.

Prevention

In 46% of downer cows the primary problem was a difficult calving. So good management at calving is vital. Good calving management is dependent upon a vast number of factors, but probably the four most important are:

— Provide a good environment: Clean, dry, low stocking density

— Ensure the cows are between BCS 2 and 3.5 at calving

— Observe from a distance, don't interfere too readily

— Know when to get help and assistance

38 percent of downer cows had milk fever as the primary cause. Preventing milk fever will significantly reduce the number of downer cows.

Cold Cow Syndrome

Cold cow syndrome is an unusual disease of unknown cause, which usually occurs in early spring in lactating cows grazing ryegrass pastures. It has been seen in most areas of the UK.

The syndrome can affect up to 80% of a group of cows The most likely cause is the intake of very high levels of soluble carbohydrates, but other suggestions include oestrogenic compounds in the plants and mycotoxins. There does not appear to be any link with weather conditions.

Clinical Signs

— Cow appears drunk, wobbling and falling over

— Cow cold to the touch, but body temperature normal

— Profuse non-smelling diarrhoea

— Sudden dramatic milk drop (up to 100% in affected cows)

Diagnosis

— On the clinical signs described above

Treatment

— There is no specific treatment, except that affected herds should be housed for at least 24 hours before they are moved to a new pasture Cows unable to stand will require nursing care

Prevention

The problem may occur on the same pasture each year, thus avoiding the use of that pasture in early spring may be helpful.

Coccidiosis in Cattle

The Nadis data show that the number of cases of coccidiosis is at its lowest in late winter, and then rises during the spring to peak in June and July. This peak is followed by a slight, short-lived fall in late summer, and then another rise to a peak in November.

Coccidiosis is caused by single-celled parasites (not bacteria) known as coccidia. There are several species in cattle, not all of which cause disease. The species that cause disease are primarily found in the large intestine, and the diarrhoea results from damage to the cells lining it. Coccidiosis is seen in animals up to two years old, and is particularly common in calves between three weeks and six months of age.

Cattle become infected when placed in environments contaminated by older cattle or other infected calves. This can happen either indoors on bedding, or outdoors around drinking or feeding troughs. In order for the coccidial oocysts to become infective they require warmth and moisture. It is probably the lack of moisture in late summer and the low temperatures in late winter that result in the low level of coccidiosis during these times, however coccidiosis can be a significant problem at any time of year.

Clinical Signs

The most common sign is a watery diarrhoea, which because the coccidia damage the large intestine is often accompanied by straining, mucous and blood. Other signs can include depression, loss of appetite, weight loss, and, much more rarely than with diarrhoea in milk-fed calves, dehydration. Death is rare. Infections that fail to produce diarrhoea can, nevertheless, result in reduced growth and weight gain. This sub-clinical infection is very common, with up to 95% of cases being of this type.

Diagnosis

— Examination of diarrhoea for the presence of large numbers of oocysts. However, care must be taken when interpreting these results and it is best to consult a veterinarian in suspect cases.

Treatment

— Most cases will recover without treatment. Discuss the necessity of treatment in particular cases with your veterinary surgeon.

— If calves become dehydrated then electrolytes should be given.

— Once high numbers of oocysts are found, then treatment is unlikely to be of any benefit

— Treatment is better given to in-contact animals that have not yet started showing signs, or to combat secondary infection. A large number of products are available for treatment, but only two are licensed. Specific recommendations should be obtained from your veterinarian.
— All calves with diarrhoea should be separated from clinically normal calves, to reduce contamination of environment with oocysts.
— If possible, during an outbreak stressful procedures, such as dehorning, castration and weaning should be avoided.

Prevention

To achieve effective control of coccidia, good management and hygiene is vital. This should include:

— Reducing stocking density
— Regularly moving feed and water troughs
— Preventing faecal contamination of feed and water troughs, by raising or covering
— Increasing the bedding to reduce contamination
— Clean and disinfect all buildings with products that kill oocysts
— Mass medication can be used as a preventative, but it is no substitute for improving management.

Botulism

Botulism is a lethal food poisoning in cattle caused by eating material that contains Clostridium botulinum toxins. The incubation period before clinical signs appear varies from a few hours to two weeks, making it difficult to identify the causative material eaten by affected animals. The most common manifestation of the disease in cattle is a subacute disease with restlessness, incoordination and difficulty to swallow developing into recumbency, paralysis and death within 1-7 days.

The bacteria and the disease occurrence are world-wide. In the UK, cases are likely to occur either due to ingestion of contaminated silage or contact with animal carcasses or skeletons of dead animals containing the toxin. The use of poultry litter as fertiliser on cattle pastures has been identified as a risk factor, due to the poultry mixed with the litter.

TETANUS

Tetanus is caused by the toxin tetanospasmin released from the spore-forming bacillus *Clostridium tetani.* The disease in cattle occurs most often after surgical intervention or difficult calving after spores gain entry to a wound. Germination of spores occurs only if the microenvironment is anaerobic. After germination of the spores within the wound the *C. tetani* bacilli proliferate and produce toxin.

The incubation period can be very variable from 3 days to several months but most cases occur usually after about 10 days. At first the animal appears slightly stiff, becomes unwilling to move and develops a fine muscle tremor. The temperature rise is variable (39 - 42°C). The general stiffness of the limbs, head, neck and tail increases after 12 - 24 hours. The animal shows hyperaestesia and repeated spasms.

Mastication becomes difficult due to tetany of the masseter muscle (lockjaw), food is chewed with difficulty, the animal drools saliva and bloat often occurs. There is retention of the urine and constipation. The animal becomes recumbent, with the legs rigidly extended, opistotonos and the jaws become rigid. The animal usually dies due to respiratory failure 3 - 4 days after the onset of clinical signs. Milder cases, which develop more slowly, can recover over a period of weeks or even months.

Control and Prevention

The most effective way of controlling clostridial diseases is by vaccination. The veterinary surgeon or the manufacturers' datasheets should be consulted for their proper use. The organic standards permit the use of vaccination in cases where there is a known disease risk. Single vaccines are preferred to more complex multiple vaccines unless such cover is specifically required.

Vaccine choice and use should be agreed with the nominated veterinary surgeon to ensure adequate disease protection during the conversion period with, where possible, progressive reductions in use as the organic unit becomes established. Only healthy animals should be vaccinated. The lower level of animal production under organic flock management will help to prevent the enterotoxaemias. Control of liver-fluke infection is an important in the prevention of bacillary haemoglobinuria.

Wounds may be treated with herbal or homeopathic remedies but it is important that wounds are exposed to air, as the organisms are anaerobes. Attempts should be made to minimise all wounds and treat all those that do occur. Good hygiene is essential at castration and assisted calving docking. Careful handling is important as bruising may cause blackleg. Clean needles and a clean, dry injection site are important if the animals are to be injected as penetrating wounds may cause tetanus.

References

Chanda, A., S. Ketan, and C.P. Horwitz., "Fe-TAML catalysts: A safe way to decontaminate an anthrax simulant," *Society of Environmental Journalists annual meeting*, Pittsburgh, October 20–24, 2004,

Meselson, M. *et al.*, "The Sverdlovsk Outbreak of 1979", *Science*, 1994.

Sternbach, G., "The History of Anthrax", *The Journal of Emergency Medicine*, 2002.

10

Dairy Waste Management

Of the total worldwide production of milk, 87% is cow milk. The rest of the production comes from buffalo (9%), sheep (2%) and goat (2%). In Europe, North and Latin America, practically only cow milk is being produced. In Asia the percentages are 58% for cow milk and 40% for buffalo milk. An important factor with respect to environmental impact is whether the produced milk is processed at home or in a factory. Home processed milk hardly offers any environmental problems as little waste is produced and as the concentration of the waste is generally low.

In developed countries, nearly all milk is industrially processed at dairy factories. A negligible proportion is used and processed at home. In developing countries this situation is completely different. In southern and eastern African countries it is estimated that about 80% - 90% of the milk is used and processed within the pastoral/agropastoral communities and their immediate vicinities. In Latin American countries this figure fluctuates between 10 and 88% with an average of 52%.

There is not a great degree of difference between the industrial methods of milk processing in developed and developing countries. It is, with respect to environmental pollution, therefore not useful to make a distinction between developed and developing countries. As will be mentioned, an important source of environmental pollution are 'house-keeping' practices that vary from country to country as well as within countries. Another important polluting factor is the production of whey during the fabrication of hard cheeses. These kind of cheeses are mainly produced in developed countries.

Developing countries hardly produce hard cheeses. Most of the production is in the form of soft cheeses or curd. These products absorb most of the whey. Dairy plants are found all over the world, but because their sizes and the types of manufactured products vary tremendously, it is hard to give general characteristics. The dairy industry can be divided into several production sectors. Each division produces wastewater of a characteristic composition, depending on the kind of product that is produced (milk, cheese, butter, milkpowder, condensate).

Milk receiving: Irrespective of the product, every factory has a section where milk is delivered and stored.

Liquid milk products: In developed countries (parts of Europe, North-America, Australia etc.), raw milk is decreamed and pasteurised or sterilised. After these steps, several products are made: consumer milk, chocolate milk, custard etc. In developing countries/regions boiling but also fermenting may be used as a means to preserve milk in the absence of refrigeration facilities. Usually, as a sanitizing method, the vessels for the storage of milk are smoked. Fermented milk may be used in fermented form but often it is churned so as to produce butter and buttermilk.

Cheese/Whey/Curd: There are about 500 varieties of cheese produced throughout the world. These are classified in nine major cheese families. These varieties come about as a result of different types of production processes. The composition of the wastewater of each specific production process varies from variety to variety. For the purpose of discussing the environmental impact, the production of cheese will be related to the production of whey. For hard cheeses, the quantity of whey produced is high and equals more or less the amount of milk used. During the production of other types of cheeses, such as soft types, the whey production is much lower or there is no production of whey at all.

Butter/Ghee: In developed countries, butter is made from cream that has been churned (separation of sweet butter and sweet buttermilk). In developing regions the technology in use for the making of butter and ghee is closely related to the technology to make fermented milk. Traditional butter is made from fully soured whole milk that is churned.

Milk powder: Milkpowder is made from raw milk, skimmilk or sweet buttermilk. After pasteurization, decreaming etc. the water from the milk is removed through evaporation.

Condensate/Cream/Khoa: For condensed milk and cream, a portion of the water is removed by evaporation. Khoa is a product typically found in India and neighbouring countries. It is produced by thermal evaporation of milk to 65-70% solid state and serves as base material for a variety of Indian sweets.

EMISSIONS

Solid Waste

Hardly any solid waste is produced by the dairy industry. The main solid waste produced by the dairy industry is the sludge resulting from wastewater purification. There are figures available about the amount of sludge production: in aerobic systems the sludge production is about 0.5 kg per kg of removed COD and in anaerobic systems about 0.1 kg per kg of removed COD.

Wastewater

Wastewater from dairy industry may originate from the following sources:

Milk receiving: Wastewater results from tank, truck and storage tank washing, pipe line washing and sanitizing. It contains milk solids, detergents, sanitizers and milk wastes.

Whole milk products: Wastewater is mainly produced during cleaning operations. Especially when different types of product are produced in a specific production unit, clean-up operations between product changes are necessary. In developing countries, the main problem is pollution through spoilage of milk.

Cheese/Whey/Curd: Waste results mainly from the production of whey, wash water, curd particles etc. Cottage cheese curd for example is more fragile than rennet curd which is used for other types of cheese. Thus the whey and wash water from cottage cheese may contain appreciably more fine curd particles than that from other cheeses. The amount of fine particles in the wash water increases if mechanical washing processes are used.

Butter/Ghee: Butter washing steps produce wash water containing buttermilk. Skim milk and buttermilk can be used to produce skimmilk powder in the factory itself or itself or these materials may be shipped to another dairy food plant by tank truck. The continuous butter production process materially reduces the potential waste load by eliminating the buttermilk production and the washing steps.

Milk powder: Environmental problems are caused by high energy consumption (= emission of CO_2, CO etc.), by cleaning and by emission of fine dust during the drying process.

Condensed milk/Cream/Khoa: Environmental problems related to the production of condensate and khoa are mainly caused by the high energy consumption during the evaporation process. The main suspended solids mentioned in the literature are coagulated milk and fine particles of cheese curd.

The production of wastewater is highly influenced by management practices. It is not possible to identify particular waste producing practices. The way in which the water consuming and operation processes are carried out is indicative of the management quality. The major contribution to he waste load comes from cleaning operations, which take place throughout the production process. Only in the production process of (hard) cheese, is whey sewering one of the main contributors to the waste load.

Waste generating processes of major significance include:

— Washing, cleaning and sanitizing of pipelines (metals), pumps, processing equipment, tanks, tank, trucks and filling machines (high N load);
— Start-up, product change over and shut down of HTST and UHT pasteurizers;
— Breaking down of equipment and breaking of packages resulting in spilling during filling operations;
— Lubrication of casers, stackers and conveyors

Air Pollution

In dairy plants air pollution is mainly caused because of the need for energy. In the process gasses may be discharged such as CO_2, CO, NO_x and SO_2. Emissions of CFC's and NH_3 into the air may come about as a result of leakage and stripping of chilling machines when out of use.

Prevention of Dairy Waste

The waste load, expressed as BOD depends to a large degree on the style of management. A large quantity of processed milk does not necessarily lead to higher waste loads or to higher levels of wastewater production. Management practices cover a wide range of water consumption and process operation activities. Well controlled processes reflect good management qualifications, while bad practices are a reflection of poor management. The qualification "fair" signifies that good as well as bad practises occur. With good management practices, values of BOD 1 kg/ton and produced wastewater below 1 kg/kg may be reached. Poor management will result in values greater than 3 kg/ton resp. 3 kg/kg.

For the evaluation of management practices, the following indicators are useful:

1. Housekeeping practices;
2. Water control practices; frequency with which hoses and other sources of water are left running when not in actual use;

3. Degree of supervision of operations contributing to either the volume of wastewater or to BOD coefficients;
4. Extent of spillage, pipe-line leaks, valve leaks and pump seals;
5. Extent of carton breakage and product damage in casing, stacking and cooler operations;
6. Practices followed during the handling of whey;
7. Practices followed in handling spilled curd particles during cottage cheese transfer and/or filling operations
8. The following of practices that reduce the amount of wash water from cottage cheese or butter operations;
9. Extent to which the plant uses procedures to segregate and recover milk solids in the form of rinses and/or products from pasteurization start-up and product change-over;
10. The procedures used to handle returned products;
11. Management attitude towards waste control.

In the following a summary is given of suggestions for the prevention of dairy waste. At the same time they are indicative of what is to be understood when speaking about good management of waste control:

1. Instruction of plant personnel concerning the proper operation and handling of dairy processing equipment. Major losses are due to poorly maintained equipment and to negligence by inadequately trained and insufficiently supervised personnel.
2. The carrying out of a study of the plant and the development of a material balance to determine where losses occur. Modification and replacement of ill-functioning equipment. Where improper maintenance is the cause of losses, a specific maintenance programme should be set up.
3. The use of adequate equipment for receiving, cooling, storing and processing of milk, so as to take care of the maximum volume of flush production and of special products. All piping, around storage tanks and other

areas, should be checked on mis-assembly and damage that may lead to leakage.

4. Accurate temperature control on plate, tubular and surface coolers to prevent freeze-on, which may result in loss of products.
5. Elimination of valves on the outlet sides of internal tubular or plate heaters and coolers and maintenance of plates and gaskets in good repair so as to eliminate waste due to blown or broken gaskets
6. Installation of suitable liquid level controls with automatic pump stops, alarms, and other devices at all points where overflows could occur (storage tanks, processing tanks, filler bowls etc).
7. Keeping in good order of vats. tanks and pipelines so as to eliminate and reduce to a minimum the number of leaky joints, gaskets, packing glands and rotary seals.
8. Proper design and installation of vats and tanks at a level high enough above the floor for easy drainage and rinsing if hand cleaned. Tanks should be pitched to insure draining.
9. Correct connections on plate type heat exchangers so as to avoid milk being pumped into the water side of the exchanger or water being pumped into the milk side.
10. Provision and use of proper drip shields on surface coolers and fillers so as to avoid that products reach the floor. Avoidance of cheese vats, vat processors or cooling tanks being overfilled so that no spillage occurs during product agitation. The liquid level in cheese vats should be at least three inches below the top-edge of the vat.
11. Avoidance of foaming of fluid dairy products, since foam readily runs over processing vats and other supply bowls and contains large amounts of solids and BOD. The use of air tight separators, proper seals on pumps and proper line connections to prevent inflow of air when lines are under partial vacuum, will avoid foam production.

12. Turning off of water hoses when not in use. Use should be made of hoses equipped with automatic shut-off valves so as to avoid excessive water usage.

References

Atkinson, D, Watson, C.A., "The environmental impact of intensive systems of animal production in the lowlands", *Animal Science: An International Journal of Fundamental and Applied Research*, 1996.

Harper, W. F. and Blaisdell, J.L., "State of the art of dairy food plantwastes and waste treatment", *Proceedings of the Second National Symposiumon Food Processing Wastes*, Denver (Colorado), USA, 23–26 March, 1971.

Harper, W. J., Blaisdell, J.L. and Grosskopf, J., "Dairy food plant wastesand waste treatment practices", US EPA, 1971.

Harper, W. J., Carawan, R. E. and Parkin, M. F., *Waste ManagementControl Handbook for Dairy Food Plants*, US EPA Publication No. 600/2-84-0, 1984.

Tess, M.W, Kolstad, B.W., "Simulation of cow-calf production systems in a range environment: I. Model development", *Journal of Animal Science*, 2000.

11

Production of Traditional Dairy Products

Milk production in India is largely based on the utilisation of crop residues like wheat/paddy straw, millet stovers etc. Since Indian agriculture continues to depend largely on the monsoon rains, the availability of crop residues is highly seasonal. Milk production in India is therefore concentrated between November—February, generally referred to as the flush season. April—September are the lean months for milk production. Two thirds of the annual milk production takes place during the four-month flush season. Milk produced in excess of the daily requirements for direct consumption was traditionally converted into various dairy products with a longer shelf life and thus the different methods of preservation of milk began.

Milk Production

The buffalo and the cow and to a very limited extent the goat are the main milch animals in the Indian sub-continent. The buffalo contributes some 64 per cent, the cow 33 per cent and the goat 3 per cent of the total milk produced in India. India has the largest milch animal population in the world. Milk of the camel, sheep and yak is used in some parts of the region. There are several well recognised breeds of cows and buffaloes in the region such as Red Sindhi, Gir, Tharparkar, and Sahiwal among the cows, and Murrah and Neeli Ravi among the buffaloes, are outstanding breeds.

Because of the lack of scientific animal husbandry and nutritional practices, the yield of the milch animals in the region has been rather low compared to that of the dairy cow in the advanced dairy countries. A beginning has been made since the middle of this century in the scientific management and feeding of cattle, highlighted by the Operation Flood programme of establishing vertically integrated cooperative dairy projects.

Milk production which amounted to 17 million metric tons in 1950–51 has increased to 46 million tons in 1986–87and crossed 65 million metric tons in 2000. The increase in milk production is to be ascribed to the establishment of rural dairies largely in the cooperative sector. These have contributed to the upgrading of animals, better feeding practices and well organised veterinary services, including artificial insemination. The population of the cross-bred cows and the upgraded buffaloes is expected to increase the milk production significantly.

The scientific organisation of Indian dairying started with the establishment of military dairy farms and cooperative milk unions throughout the country towards the end of the nineteenth century. Organised marketing with the application of advanced dairy technology commenced only in 1954 with the establishment of the Amul Dairy which was the first dairy in the region to manufacture milk powder, condensed milk and cheese from buffalo milk. Amul Dairy now handles some one million litres of milk per day.

Conditions and Hygienes for Milking

Most of the milk in India and the neighbouring countries is produced in the villages by farmers with small land holdings and also by landless agricultural labourers. Although an increasing portion of the milk produced is collected by the cooperatives and other organised dairies, a significant portion of the milk is still being converted into traditional dairy products due to lack of refrigeration and transportation facilities. Conditions under which milk is produced in the villages are far from satisfactory, mainly because of the economic backwardness of the producers.

The milk animals are housed in a part of the living space of the family or in small closed or open yards adjacent to the family house. Flooring is usually a plaster of mud. The cows are rarely washed before milking. Buffaloes generally wallow in ponds, especially in the hot summer months. Milking is done by hand, usually after suckling by the calf. Except in a few modern large farms, milking machines are not used. Because of the distances between the producing and consuming points, milk is unavoidably held at ambient temperatures for a significantly long time leading to high microbial growth.

The high ambient temperatures in the region for the major part of the year support rapid microbial growth. The predominant types of microflora in milk received in dairies are coliforms, micrococci, lactic streptococci, spore-forming aerobes and corynebacteria, the majority of these being contaminants from milk utensils. There is also a high incidence of thermoduric bacteria.

Production Levels

Milk is mostly produced in small quantities, of 2–4 litres, by small and marginal farmers in numerous and widely scattered villages. The farmers whose principal occupation is agriculture, keep a few cows or buffaloes for milk production (2–4 animals on an average) as a supplementary source of income. The collection, transport and distribution of fluid milk under the tropical conditions prevailing in India and the neighbouring countries present many difficult problems.

The production of milk in villages takes place on a very small scale in numerous scattered holdings, which makes the task of collection difficult. Many villages are not connected by good roads, and many more are inaccessible during the monsoon rains. There are no facilities for cooling or refrigeration of milk on receipt at a village collection centre and rapid transport to a processing centre is hampered by lack of facilities and infrastructure. Under these conditions. procurement of milk of suitable quality in a condition fit for processing into marketable products is a formidable organisational task which has been performed well by many dairy cooperatives on a fairly large scale.

Numerous agencies and persons are involved in the collection, transport and distribution of milk; village producers who directly supply milk to village cooperatives, milk collectors who collect milk from producers and supply to the collection centres of organised dairies or to urban areas or halwais, milk vendors, dairies who process market milk and milk products, wholesalers and retailers. Milk may be carried to the collection point as headloads, or in containers suspended over shoulder slings, on bicycles, on pack animals, or horse drawn carriages depending upon the quantity of milk to be transported and distances involved.

In the case of villages situated at greater distances, the milk collectors or agents may transport milk cans in trucks or by rail for sale directly to consumers or halwais and to collection centres or private dairies. In some cases, milk is also transported in small boats. For long distance transport of milk, galvanized iron cans or factory made aluminium alloy milk cans are generally employed, and where the collection and transport of milk supplies are organised by milk cooperatives or by large dairies, the cans are cleaned and steamed at the dairy and then returned to the collection centres. Public sector/cooperative dairies have their own milk collection and distribution systems and their major responsibility is that of distribution of pasteurised milk to consumers.

These dairies now account for some 25 per cent of all the milk marketed in India. Cooperative milk unions have been organised in several parts of the country and have tackled successfully the twin problems of marketing the rurally produced milk and of supplying good quality milk to the urban population. Presently, some 447 cities in India are served by the organised sector, handling over some 7 million litres of hygienically processed milk per day. They cater for some 55 million customers.

The demand for fluid milk of the four major cities in India (Bombay, Calcutta, Delhi and Madras) is estimated at about 6 million litres per day of which nearly 50 per cent, or 3 million litres per day, is provided by the ten public sector dairy plants in these cities. The remaining 50 per cent of the demand is still being supplied by the traditional milk trade. Milk produced in large scale organised farms with herds of from 100–1000 head of cattle and buffalo constitutes only a negligible fraction of market milk in India. Some of these farms also have pasteurising plants and cold storage facilities. Some 10 per cent of the milk produced in India is processed by the organised sector in 250 dairy plants in the cooperative, public and private sectors with a combined throughput of 12 million litres per day.

Milk Production Technologies

It is evident from the contributions to this review that in most of the countries to which reference is made, that conservation and prevention of the spoilage of milk remain of major importance. The producers of milk in the countries under review have not been influenced to a major extent by improvements in refrigeration and milk collection methods such as have had a major effect on dairying practices, milk conservation and milk quality in the past thirty years in countries with a developed dairy industry.

Methods to protect milk from spoilage remain empirical and yet they appear to offer at least some help to the milk producer. In reviewing the technologies employed in the preparation of traditional milk products it is necessary to appreciate that the characteristics and quality of the milk produced in the countries concerned is not comparable in bacterial content with the milk from dairy farms in developed dairy countries which are equipped with refrigerated milk coolers and bulk milk collection.

Nevertheless, practices for the limiting of spoilage of milk are in place in the countries concerned and include immediate boiling of milk after its production, the use of lactose fermentation by lactic acid bacteria as a means of preventing milk spoilage and sanitising methods which include smoking of the vessels used for milk and milk products. It is mentioned above that fermentation of lactose to form lactic acid is an important means of preventing, or limiting, milk spoilage due to the growth of contaminating bacteria and their enzymic activity.

The value of lactic fermentation as a means of preservation has led to a situation in Africa, the Middle East, and India and neighbouring countries, that many of the processes for traditional milk products include a fermentation stage. This stage not only affects the shelf-life of the product but it also affects the quality and characteristics of the product.

As an example, the souring of milk leads to the preparation of products such as dahi and other highly soured milks which can either be consumed in liquid form or may be processed further with separation of the milk constituents into traditional butter which is the basis of ghee, and soured milk which may be eaten as it is or be further developed into various cheese products. It is clear that very many variations in the characteristics, quality and acceptability of traditional milk products are inevitable due to the unregulated nature of natural fermentations which have such an importance. In many cases little is known of the exact nature of the bacteria or other micro-organisms contributing to these fermentation processes.

Fermented Milks

Dahi

This milk product is of major importance in the Indian sub-continent. It is a yoghurt-like product made in India and neighbouring countries. It is the most important fermented milk product used in India from times immemorial. The scale of production ranges from household level to industrial scale including preparation in halwai's milk shops in urban areas. Cow or buffalo milk or a mixture of the two is used. It is boiled and sometimes concentrated before addition of the starter which is usually a portion of the previous day's dahi or buttermilk.

Dahi has a milk pleasant flavour and a clean acid taste. It has a yellowish creamy-white colour when made from cow milk and a creamy-white colour when made from buffalo milk. It has a smooth and glossy surface. The body is firm but not hard and free from gas holes. Dahi is widely consumed all over India and the neighbouring countries including the Himalayan region, either plain, sugared or salted. The sweetened concentrated form of dahi consumed in Bengal is known as mishti doi i.e. sweet dahi.

Mishti Doi

A sweetened variety of dahi known as mishti doi, mishti dahi, lal dahi (red dahi) or payodhi in the eastern region of the Indian sub-continent is very popular. Cane sugar (6.0 – 6.5%) is added to the milk before boiling. Artificial colour, caramel and jaggery may also be added. The milk is cooled to 40–45°C and incubated for 12 – 15 hours.

Lassi

Dahi is converted into this refreshing beverage by stirring and adding a small quantity of water. It is best consumed chilled, and either sweetened or salted. It is a preferred drink in the northern parts of the sub-continent particularly the Punjab and Haryana. It is known to induce sleep particularly after consumption during the summer afternoons. Aseptically packed long life lassi has recently been introduced in India.

Shrikhand

Shrikhand or Sikarni, as it is known in Nepal, is made from concentrated dahi with a sweet and sour taste. It is a semi-soft whole milk product resembling sweetened quarg or quark produced in Germany. Shrikhand is traditionally made at home in western India. The name shrikhand is derived from the Sanskrit work shikharini. Dahi is placed in a muslin cloth and drained for 4–8 hours to reduce the whey content and produce a solid mass called chakka or maska. Chakka is mixed with the required amount of sugar, condiments and flavour to produce shrikhand. An industrial process for the manufacture of chakka and shrikhand has been developed by the National Dairy Development Board of India.

Shrikhand Wadi

This product is obtained by further concentration of shrikhand as prepared above by heating in an open pan over a direct fire until it forms a hard mass.

Shrikhand wadi has the following composition.

Moisture	5–6(per cent)
Fat	7–8
Protein	8–10
Lactose	15–17
Ash	0.75–0.80
Sugar	63–65
Lactic acid	1.0–1.2

Chhaas (Buttermilk).

Buttermilk produced by the churning of soured milk (dahi) is known as chhaas or chhach and as mahi in Nepal. The fat content is usually from 1–2 per cent and it is rich in protein and lactose. Chhaas is mostly consumed in the household and surplus is fed to cattle.

Kadhi.

This product is made from chhaas by a recipe which varies from region to region. A blend of spices, of which the common ingredients are salt, black pepper, green chillies, turmeric, coconut, and ground cumin are added with a small amount of Bengal gram flour to an appropriate quantity of chhaas and the mixture is brought to boiling point. It is then served hot with rice. In some regions of the country, small balls made out of besan dough and fried in oil are added to kadhi and served as a curry.

Butter and Ghee and Related Products

Makkhan

Makkhan is the traditional unsalted butter made by hand churning whole milk dahi. Beginning from the vedic times there is recorded evidence to show that makkhan was extensively used by the early inhabitants of India; both in dietary and religious practices. The milk of the water buffalo, by virtue of its higher fat content and larger fat globules gives higher yields and is preferred. White makkhan from buffalo milk is generally preferred to the yellower product from cow milk. About 170,000 metric tons are estimated to be produced annually in Indian households.

Makkhan is used in small quantities for direct consumption with the traditional unleavened bread or boiled rice and other items of food. Household surplus of makkhan is used mainly for conversion into ghee. Organised dairies, produce butter on a commercial scale using modern butter making machines. Only a part of the butter produced in India is used as table butter and a large portion is used for the production of ghee on a commercial scale.

Ghee made from makkhan has a firmer consistency, better crystalline texture and reputedly a better shelf-life than the product made in factories. Whenever boiling of milk is carried out over a smoky fire, the makkhan produced from milk heated in these conditions has a typical smoky flavour which is often preferred by a section of the consumers, particularly in the northern region.

Cheese Products

Paneer

Paneer consists mainly of acid-coagulated milk solids and is used extensively as an ingredient in many cooked vegetable preparations in Northern India, Pakistan, Afghanistan and Nepal. Paneer making is confined to the North-west frontier regions of the Indian sub-continent. It is produced at small scale and industrial level. Cow, buffalo or mixed milk may be used but buffalo milk is preferred.

The milk is boiled and the coagulation is simultaneously effected by adding the required amount of coagulant acid in a thin stream, within a minute, and mixing it into the milk with a stirrer. Draining is begun when the whey is clear. On a commercial scale, Paneer is processed mechanically into blocks in hoops by putting weights on the hoops (approx 2–3 kg per sq cm for 15–20 min). Drained and pressed curd is cut into suitable sizes and immersed in chilled water for 3–4 hours to make it firm. It is usually sold in pieces without packaging.

An industrial-scale process has been developed by the NDDB. Milk is heated to 85°C through a plate heat exchanger and pumped to a cheese vat and cooled to 75°C. Citric acid solution is added and mixed with the milk to form a coagulum. The curd is left to settle for 10–15 min without agitation. The whey is drained off. Curd is filled into cheese hoops lined with muslin cloth. Pressing of the curd for 10–15 min at a pressure of 3 kg per sq cm. Pressed curd blocks are place in pasteurized cold water at 4°C for 3 hours. The cooled blocks of paneer are cut into 200 g or 500 g portions which are wrapped in vegetable parchment paper before being placed in HDPE or LDPE bags and heat sealed ready for sale.

In India, paneer must meet the following legal requirements:

- The yield of paneer depends on the quality of milk. It is generally 18 to 20 per cent of the weight of the milk used for its preparation.
- 'White' paneer is a staple food of nomads in Afghanistan. It is traditionally consumed in the northern regions of the Indian sub-continent with dry fruits and nuts as a dessert.
- Paneer is also the Hindu name of the seeds of Withania coagulans, the basis of a vegetable coagulant that yields a bitter curd.

— Curdled milk products obtained by the admixture with sour milk, pieces of a creeper called Putika, the bark of Palasa trees or Kuyala (Jukuke) was known to the ancient Indians.

However the curdled milk product, paneer, seems to have been introduced into India from the Middle East perhaps by Persian and Afghan invaders. A unique Iranian nomadic cheese is called paneer Khiki. This cheese was originally developed by the well-known Bakhtiari tribe which resided in Isfahan (in summer) and Shiraz (in winter). The word khiki means skim.

Rennet from the goat or sheep was used to make the paneer, hence the name. When salted it known as paneer-e-shour. It is only in the past four decades that consumption of paneer has spread to other parts of India. It enjoys the status of haute cuisine amongst Indian vegetarian cooking.

Surti Paneer

The name of this cheese is derived from the town of Surat in western India where it was probably first prepared and marketed. Once a popular product, very little of it is marketed today. It is a soft cheese prepared from buffalo milk with crude rennet, salted and kept steeped in acid whey for 2–3 days. Surti paneer should have a fairly firm body and smooth texture with no internal cracks. It has a slightly salted, milk acid-curd flavour.

Bandel Cheese

Bandel cheese is an indigenous unripened, salted soft variety of cheese made in perforated pots. It is similar to surti paneer but made from cow's milk. It is available in and around Bandel, a Portugese colony in eastern India, and seems to have derived its name from it. The cheese is formed into a flattened circular shape and is ready for immediate sale.

Dacca

This cheese is available in the eastern region. It is similar to bandel but differs from it in that the finished flat round cheeses are smoked in a fire.

Chhanna

It consists of acid coagulated milk solids used for the preparation of many milk based sweets. It differs from Paneer in that no pressure is applied to remove the whey. Chhanna is widely used in the eastern parts of India and Bangladesh. Cow milk is preferred since it yields a soft bodied and smooth textured product. Both these characteristics are suitable for the production of high grade chhanna sweets.

Buffalo milk produces a chhanna with a slightly hard body, a greasy and coarse texture, and does not produce good quality chhanna sweets. Chhanna has the same legal requirements as paneer in India, i.e. a maximum moisture content of 70 per cent and a minimum content of milk fat in dry matter of 50 per cent.

Chhanna from cow milk is light yellow in colour, has a moist surface, soft body and smooth texture. Chhanna derived from buffalo milk is whitish in colour. Both have a pleasant sweetish, mildly acid taste. Buffalo milk yields a larger amount of chhanna. About 100,000 metric tons are produced annually in India. Chhanna is also produced in rural milk sheds and transported by road and rail to larger urban conglomerates in wicker baskets which allow further drainage of whey. Chhanna produced in this way is used for the preparation of Sandesh.

Chhanna-based Sweets: Rasogolla

This sweet is of recent origin having been developed in 1868 by an enterprising Calcutta sweetmeat maker Nobin Chandra Das. It is prepared using fresh and soft-chhanna. In the form of balls 30 mm in diameter with a typical spongy body and smooth texture. Stored and served in sugar syrup. Freshly-made chhanna is squeezed by hand in a muslin cloth to remove as much whey as possible. 1–4 per cent of the wheat flour/semolina is mixed with the chhanna in a container and kneaded thoroughly by hand to make a dough.

The dough is portioned and rolled into balls of about 15 mm diameter having a smooth surface with no cracks—1 kg of chhanna yields 90–100 rasogollas. The dough balls are cooked in a specially prepared whey based medium for about 15 minutes. For chhanna made from cow milk, cooking medium with sugar is preferred, and for all other types of chhanna, cooking medium without sugar is preferred.

After the cooking is complete, the balls are transferred to a container with water at 30–35°C for texture stabilisation and colour improvement of the balls. After 5–10 min of texture stabilisation in water, the texture stabilised balls are transferred to sugar syrup. The desired sugar syrup concentration in the final product is 45–50 per cent. This is achieved by dipping the texture-stabilised balls first in 35–40 per cent sugar syrup for 1–2 hours, followed by a second dipping in 58–60 per cent sugar syrup. The product finally acquires the desired sugar concentration after equilibration between the sugar syrup inside and outside the balls is achieved.

Lalmohan

A product similar to gulabjamun but is made from chhanna and is lighter in colour. Chhanna is mixed with 2–3 per cent wheat flour and kneaded into a uniform dough. The dough is rolled into small balls and deep fried in ghee until light brown in colour. The balls are transferred to a 60 per cent sugar syrup and allowed to soak for a few hours before being served.

Other Milk-Based Products

Khoa

This product is obtained from cow, buffalo or mixed milk by thermal evaporation of milk to 65–70 per cent solids in an open pan. A five times concentration of milk is normally required for the production of khoa. Khoa, also khawa or mawa, is used as a base material for a variety of Indian sweets. Its origin is not known but it has been prepared for centuries in India as the base material for sweets. About 600,000 metric tons of khoa is produced annually in India alone. It is made by the traditional method by milk traders and halwais.

Khoa preparation has been the easiest way of preserving rurally-produced milk in the flush season. In many places khoa manufactured in January—February is cold-stored for use in the summer season. Such khoa acquires a green colour due to mould growth on the surface. It is therefore known as hariyali (green khoa). This khoa is preferred for the preparation of gulabjamun as it gives a grainier texture to the product. This khoa, on removal from the cold store is immediately mixed with flour and made into gulabjamuns. Hariyali khoa, if left at room temperature for long, starts to smell and breaks down physically. Because of this it is converted into products immediately.

Legal requirements state that khoa contains a minimum of 20 per cent milk fat. The Bureau of Indian Standards has laid down the following specifications for khoa. Milk of high acidity produces a granular khoa known as danedar. Khoa has a uniform whitish colour with just a tinge of brown, a slightly oily or granular texture, and a rich nutty flavour which is associated with a mildly cooked and sweet taste due to the high concentration of lactose.

Buffalo milk is preferred for khoa making because it yields a whiter product with a soft, loose body and a smooth granular texture which makes it suitable for the preparation of high-grade khoa sweets. A minimum of 4 per cent fat for cow milk and 5 per cent fat for buffalo milk is necessary to obtain a desirable body and texture in khoa. Lower levels of fat result in undesirable hard body and coarse texture. The traditional trade usually pays for milk on the basis of the yield of khoa. Cow milk usually yields 18 per cent of khoa. The yield from buffalo milk is usually 20 per cent.

Peda

The quantity of peda produced in India exceeds any other indigenous milk-based sweet using khoa as the raw material. Peda or doodh peda is prepared on a small scale by halwais using khoa as the base material mixed with sugar and flavourings. A similar product which is very popular in Nepal is called gundpak.

Peda is usually packed in paperboard cartons with a parchment paper of grease proof paper liner. it is usually sold through confectionery shops Peda is whitish yellow in colour and has a coarse grainy texture. Kesar (saffron) peda is one of the preferred pedas in which saffron is used for added flavour and colour.

Other Khoa-based Sweets.

The methods of preparation and various features of other khoa-based sweets—burfi, kalakand, gulabjamun, and kalajanum or kalajam.

Condensed Milk-based Products

This sub-group of milk-based products includes rabri, khurchan, basundi, kheer and palpayasam. Features of preparation are given below.

Rabri, Tar (in Nepal): A specially prepared concentrated and sweetened whole milk product containing several layers of clotted cream. It is a sweet by itself and is not much used as a component of other sweets. It is produced in Northern and eastern regions of India normally from buffalo milk is normally used since it produces a more creamy and chewy consistency. In comparison to cow milk the higher fat and casein contents of buffalo milk contribute to the formation of a greater volume of creamy layer early in the evaporation process.

Milk (3–4 kg) is heated in a fairly shallow pan over an open fire and allowed to simmer, 5–6 per cent of sugar is added and evaporated to one eighth of the original volume. The preparation time is about 25–40°C minutes depending on the rate of boiling. The finished product consists of non homogeneous flakes partly covered by and partly floating in sweetened condensed milk. By heating the concentrate slightly at the end, a more homogeneous chewy-textured mass is obtained. The following composition relates to Rabbri prepared in Nepal.

Khurchan: A concentrated, sweetened whole milk product, similar to Rabri. It is used for direct consumption. It is produced in the Northern region of India almost exclusively from bufflo milk as it gives a higher yield than cow milk. The final produce has a slightly cooked flavour, which is relished.

Basundi: A concentrated milk to which sugar, flavours and nuts are added. The product is served chilled as a dessert. The origin of the product is not known but it has been traditionally prepared for centuries in the western part of India as a dessert, served on special occasions such as weddings. About 25,000 metric tons of basundi are produced annually in India from cow and buffalo milk on a small-scale.

Milk, in a shallow pan is boiled on a low flame. The heat coagulated film that appears on the surface of the milk is collected and spread on the sides of the vessel. The volume of milk is reduced to 50 per cent of it's original volume. The pan is removed from the fire and sugar is added along with nuts and flavours. The mass is mixed until the sugar is dissolved. The product is cooled and served chilled. Basundi looks like condensed milk with flakes. It has a light brown colour with thin flakes in a thick fluid. It has a pleasant flavour similar to condensed milk. The cooked flavour is relished by the consumer.

Kheer: A sweetened product of thick consistency resembling rice pudding commonly consumed in the West. The product is prepared for immediate consumption. It is produced in northern, western and central regions of the Indian sub-continent. This product is widely consumed in the regions mentioned above. In Nepal, the housewife perpares kheer by concentrating whole milk in open pans with the addition of sugar (6–8 per cent), rice (6–7 per cent), ghyu, cashew nuts, cardamon and other spices.

Palpayasam: A sweetened product; similar to kheer and resembling a rice pudding produced in Southern India. Vermicelli or semolina may be substituted for rice, fruits like jack fruit are optional.

Frozen Milk Products: Kulfi or Malai Kulfi.

This indigenous ice cream is based on milk and is popular in the hot summer. It is frozen in small containers. The preparation of kulfi involves concentration of a milk and sugar mixture to 50 per cent volume. It is cooled before addition of cooled cream, crushed nutes and selected flavourings. The milk is added to moulds and frozen in a vessel containing an ice and salt mixture with a 1:1 ratio. The Bureau of Indian Standards has laid down the following specifications for kulfi.

Ecconomic Importance of Milk Products

With an annual milk production of 46 million metric tons in 1986–87, India ranks third in the world after the Soviet Union and the United States of America. Milk production is expected to further surge forward in the coming years to cross the 61 million metric tons mark by 1995. The target for 2000 AD is 70 million metric tons. Milk and milk products play a vital role in the agricultural economy, being the second largest agricultural product in India.

In 1984–85 the value of milk and its products exceeded Rs.100,000 million, ranking after rice, but before wheat. Equally important, if not more so, is the role of dairying in providing sustenance to millions of farmers, constituting 75 per cent of the total population in some 80 million farm households distributed over 550,000 villages, constituting the bulk of rural poor, with an annual income of less than Rs. 3,800 per family. Milk provides both nutrition and supplementary income to these weaker sections.

Over 5 million farm families, in 49,000 village milk producer's cooperatives, sell on an average some 8 million litres of milk every day, after retaining some 30 per cent of it for their own consumption. Sale of milk fetches them about Rs. 30 million per day aggregating Rs. 10,000 million per year. India alone produces some 650,000 metric tons of ghee valued at Rs. 32,000 million. The value of resultant lassi is Rs. 10,000 million, some 600,000 tonnes of khoa valued at Rs. 18,000 million is produced in India along with some 100,000 metric tons of chhanna valued at Rs. 3,000 million.

The value of khoa and chhanna produced in India is probably twice the value of all the milk handled by the organised sector in the country. The traditional dairy products sector in India, like its agricultural economy, is grossly under managed. However, it provides economic opportunities that even the western dairy world would be envious of. The value of khoa and chhanna-based sweets could possibly exceed US $ 4 billion. In the absence of reliable data, the above figures are only rough estimates, but highlight the significance of the traditional dairy products in the national economy.

Nutritional Importance of Milk Products

The composition of milk makes it an ideal balanced food for humans especially infants and its importance as a supplement to the average diet cannot be over emphasised. Traditionally in areas where milk production is abundant, milk and milk products are regularly consumed by almost all sections of the population. For example, the average Punjabi diet can compare well with some of the best diets in the world. However, the same cannot be said of the major sections of the population.

Recognising the proper role that milk can play in the nutrition of the people, efforts are being made to increase milk production significantly. Supplementary feeding programmes for infants and expectant mothers, and school children, have always included milk powder as one of the ingredients. There is hardly any major difference in the nutritive values of cow and buffalo milk except for the greater calorific value of buffalo milk due to its higher fat content.

Although 46 per cent of the milk produced in the country is consumed as liquid milk and as such milk plays an important role in the national diet, there is considerable need and scope for increased consumption of milk. The expenditure elasticity of demand for milk is very high in India, 1.46 for the rural population and 1.3 for the urban population.

The daily allowances of nutrients for an Indian adult male recommended by the Indian Council of Medical Research are given below. The average diet of the poorer sections of the population is deficient in several nutrients and most of these can be made up by supplementing the diet with milk. As against the recommended level of 200 ml of milk, the average per capita intake was 168 ml, in 1988. Except in the case of high and middle income groups it is less than the recommended levels.

Milk plays a major role as a source of proteins in the average Indian diet contributing some 10 per cent of the protein intake. These data are indicative of the important part that milk plays in the nutrition of the population. In India most milk is boiled before consumption. It is to this practice that the absence of milk-borne diseases in India is to be mainly attributed. Heating to first boil results in destroying most of the organisms. Denaturation of proteins as well as its flocculation due to the neutralisation of the electric charges occurs to some extent on boiling milk.

A partial precipitation of calcium salts and phosphates also occurs, the diffusible calcium being reduced from 26 per cent to 20 per cent. Among the vitamins in milk, A is the most resistant, and C the most vulnerable to heat treatment. While hardly any vitamin A is destroyed by boiling, about 22 per cent of vitamin C is lost when milk is boiled. The loss of vitamin C is dependant both on the time of treatment and the exposure to light. A slight reduction in the thiamine (B1) content of milk occurs. Riboflavin (B2) is hardly affected. The availability of calcium and vitamins (except vitamin C) is not affected by boiling.

Most of the enzymes of milk are destroyed during boiling and the digestibility of milk increases. Hot milk is widely consumed before going to bed as a nightcap. The milk is usually flavoured with condiments such as almonds, cardamom, dry dates etc. In the Indian households the life of milk is extended from 12 to 24 hrs by repeated boiling. The simplest way of preserving milk for human consumption in a tropical country is to allow it to sour with the aid of lactic cultures, checking putrefactive changes while giving to milk an acid taste which is particularly refreshing in a hot climate.

The product thus achieved, dahi, is widely consumed in the country along with meals. The digestibility of milk constituents improves. Dahi can also be consumed by people who suffer from lactose intolerance. Almost every household in the country consumes dahi. Due to fermentation of milk a greater amount of phosphorus and calcium is made available to the digestive system by their precipitation in the lower intestines due to the acid condition induced by Lactobacillus sp.; and the consumption of sour milk also results in increased efficiency of the body to cope with a sudden influx of lactic acid in the system.

It is reported that when the food is supplemented with 250 g of dahi a day, the status of thiamine improves. Dahi also increases the pyruvic acid and the lactic acid among children on a typical poor rice diet. Thus, dahi in its different forms, lassi, kadhi, shrikhand etc. also contributes significantly to the average diet. Makkhan and ghee contribute as much as one third of the fat in the Indian diet.

Ghee is produced mainly for consumption directly as food and as an ingredient of food preparations including sweets. Over the centuries Indians have cultivated a liking for the aroma and flavour of ghee, and a preference for its use over vegetable oils, the other traditional cooking medium for the preparation of specific food items.

The vegetarian habits of many Indians preclude from their diet hard animal fats such as tallow or lard used in the West and thus ghee forms an important source of fat in an otherwise vegetarian diet. For most uses, its wholesome flavour is the chief attraction. For table use it is served in melted form and mixed with rice or lightly smeared on chapatis. It is widely used for shallow frying and deep frying of food materials. Innumerable Indian sweetmeats based on cereals, milk solids, fruits and vegetables are cooked, by preference, in ghee.

Buttermilk or lassi as described earlier is a by-product in the preparation of makkhan. It is estimated that about 55 kg of buttermilk is produced for every kg of ghee. While most of this is consumed by the villagers and their families, a good quantity is either given away or fed to cattle. The reason for this is the lack of market value for the product in rural parts. Buttermilk is rich in milk protein and calcium and forms a valuable human food.

Ghee and makkhan are important carriers of vitamins A, D, E & K. They also contain small amounts of essential fatty acids e.g. arachidonic and linoleic. Considerable losses of Vitamin A and carotene occur during cooking, the loss of the latter being more rapid. Below 125°C Vitamin A is fairly stable but above this temperature it is rapidly destroyed. It is found that 10–20 per cent of carotene is lost during the normal cooking operations.

Marketing of the Dairy Products

Marketing of the indigenous dairy products is as traditional as are the products. Halwais, the traditional sweetmeat sellers, produce and sell these products in all urban and semi-urban areas of the region. Halwais have prospered over the years as the products have high margins of profits.

The festival season (October—November) sales in many areas account for 30–40 per cent of the annual sales of traditional milk-based sweets. Most sales are made across the counter for ready cash. Kulfi (ice cream) is usually sold from door to door by hawkers and some manufacturers also offer kulfi as a part of their wide range of ice creams.

Ghee is usually marketed in the traditional markets (mandies) by those who collect it from the villages and refine it further to remove all moisture. From the mandies, it goes to retail outlets. Ghee is usually branded by large traders. Large ghee mandies exist in Hathras, Khurja, Porbandar, Guntur and Erode. Jodhpur is probably the largest ghee trading centre in India. Some manufacturers now advertise ghee on the national television network. Organised dairies brands of ghee now fetch a good price as the quality is considered to be guaranteed.

Expanding markets: Some of the traditional halwais have more than one outlet in a city. Some have started canning rasogollas and gulabjamuns. Bikaner in the west has emerged as a large custom processing and manufacturing centre for rasogollas. Halwais in Bikaner execute large orders and provide custom labelling for canned rasogollas which are manufactured in the traditional way but do not reach the quality of the traditional product.

Modernisation of marketing efforts: The most modern plant manufacturing traditional dairy products is the Sugam cooperative dairy at Baroda marketing its products through a large network of 150 retail outlets in the city. The Sugam Dairy uses the traditional grocery/ general stores that have a refrigerator to market its products. The product range includes shrikhand, gulabjamuns, peda and lassi apart from flavoured milks. The dairy has the highest turnover of a single unit marketing traditional dairy products. Other dairies are active in marketing traditional milk products.

References

Morr C. V., "Whey and Whey Products Available for Utilisation, in Wheyand Lactose Processing", edited by Zadow, J.G. *Elsevier Applied Science*, London, 1992.

Robinson, R. K., *Modern Dairy Technology*, Volume 1. Elsevier AppliedScience Publishers, London, 1986.

Varnam, A. H. and Sutherland, J. P., *Milk and Milk Products:Technology, Chemistry and Microbiology*, Chapman and Hall, London, World Bank, 1997.

12

Small-scale Dairy Business

With increasing population and urbanisation, demand for marketed milk and dairy products was created by households who did not have the time, land or inclination to produce themselves. Initially traded through barter exchange with other household necessities, milk and dairy products are now commonly sold. Although in some countries a large proportion is still sold directly to consumers, distance to major markets resulted in the need for middlemen to link producers and consumers.

The demand for preserved dairy products spurred commercialisation and processing of milk into higher value products like cheese, yoghurt and butter. The employment generated by small-scale dairy marketing and processing in the three cases presented here is associated with procurement, transportation and sale. For the purposes of this study the dairy traders considered are small scale, handling a maximum of 5 000 litres of milk daily. Within this group of traders, those who handle up to 500 litres are categorised as very small traders. Processors are those who either pasteurise milk or produce other added value products such as cheese and yoghurt for sale.

Types of Small-scale Dairy Marketing

Dairy product markets typically differ in several key ways: a) by the types of products handled, and b) by the number of intermediaries involved, and the role each plays. These two aspects are often linked in that more processed and thus higher value products often involve more intermediaries, each of whom adds some delivery or transformation service to the product. Simple distance between source and sales areas, or the density and scale of the production system, even without product transformation, can however also increase the number of intermediaries, due to the need for assembling, bulking, transporting and distributing.

The cases of Kenya, Ghana and Bangladesh represent increasing product differentiation, but each has its own complexity. In Kenya, the market is mainly for liquid milk, but distances require relatively complex market channels and often several intermediaries. In Ghana, there is more product differentiation, now with markets for local cheese as well as milk. In Bangladesh, there is a wide range of products, from sweets to hot milk to sweetened curd, requiring a similarly wide range of intermediaries.

The three countries also differ in the quantities of milk and dairy products consumed. Kenya, with a strong indigenous tradition of milk consumption, has a very high level of consumption, well above the average of 55 kg for Sub-Saharan Africa. In contrast, Ghana exhibits very low levels of consumption, in keeping with the absence of a dairy culture among most of its population, particularly those in the southern parts of the country, where trypanosomiasis has limited cattle keeping.

Bangladesh has lower levels of dairy product consumption than other countries in South Asia, particularly Pakistan. Curiously, it is clear that higher levels of consumption are not necessarily associated with consumption of more concentrated dairy products; the consumption in Kenya is mostly of liquid milk in tea. The movement of milk in Kenya bears a close resemblance to that in Bangladesh in that producers are not much involved beyond production and farmgate sale of the milk.

In Ghana, stockmen and their families, however, also play a major role in milk processing and the sale of processed milk products. The other major difference is that whereas the milk market in Kenya is mostly of liquid milk, processed products play a major part in dairy marketing and consumption in Ghana and Bangladesh. Special cases of middlemen in Bangladesh who facilitate transactions of larger quantities than 5 000 litres of milk on commission are included because they play a major role in facilitating the small-scale milk trade.

Kenya

Dairy farming with improved (exotic) cattle in Kenya dates back to the beginning of the century, before which local cattle were kept and milked. Until the early 1960's, formal dairy marketing and processing were predominantly in the hands of large-scale settler farmers and a parastatal, the Kenya Cooperative Creameries (KCC), with focus on butter export markets. After independence in 1963, improved dairy cattle increasingly came into the hands of small African farmers as settlers sold their herds.

Small-scale dairy processing and marketing in Kenya received a further boost following policy changes that liberalised dairy marketing in 1992, and improved the prices they received for milk. Since the policy shift, the dairy industry has seen major changes ranging from the near-collapse of the KCC, to the emergence of many small and large-scale milk traders who participate in both formal and informal milk markets.

The dominant raw milk markets currently employ thousands of people who are estimated to handle about one-third of the total marketed milk produced by some 600 000 small-scale farmers. It should be noted that some 55 percent of all milk marketed is sold directly from farmers to neighbouring consumers and institutions. Only some 8 percent of milk is sold directly to processors, with small traders and cooperatives handling the rest, some 38 percent.

The small milk traders comprise those who trade in milk as the main business (mobile/itinerant traders, milk bars and processors) and those who trade in milk as part of other retail activity mainly involving sale of other household consumer items (shops/kiosks). In the latter case, the milk trade often comprises less than half of total turnover.

Mobile Milk Traders in Kenya

Mobile milk traders in Kenya are characterised by the absence of fixed business premises. The proprietors predominantly run their businesses personally. Their mode of transport is mainly by bicycle but also includes public transport and on foot. The majority typically sells 50-120 litres of raw milk daily. The milk is collected from rural areas up to 150 km away, but more often 30 to 60 km distant, and delivered to customers in urban areas. They enjoy the highest returns to labour per litre of milk sold compared to other traders, up to four times higher than those do by other traders.

A major current constraint in the mobile milk trade is the lack of legal recognition of the trade by Kenya Dairy Board (KDB) regulators who argue that the lack of fixed premises compromises milk quality. The mobile traders therefore do not qualify for trade licences from the KDB. However, they have tried to look for ways to circumvent the requirements of the KBD.

Milk Bars and Shops

Traders with milk bars have fixed premises and mainly sell both unpasteurised and fermented liquid milk. They vary from those with a sitting arrangement for customers to those that act only as retail or wholesale/distribution points. Fresh and fermented milk and sometimes yoghurt are sold. Besides family labour, wage employment is actively involved in running the milk bars. Milk collection from producers is mainly on foot, by bicycle or public transport. Own vehicles are used in a few cases.

Small Processors in Kenya

Small processors in Kenya mostly process and sell pasteurized milk, with a small proportion of throughput devoted to yoghurt and cheese, either as wholesalers and/or retailers.

They are much fewer in proportion to other cadres of milk traders. The majority of small processing enterprises have only recently started the business, but a number have been in operation for several decades. The activities requiring labour in milk processing enterprises comprise collection and processing of the value added products, packaging, distribution and sale. Processing into the value added products requires training. However most workers employed in dairy processing are mainly trained on the job and the majority are male.

Bangladesh

Bangladesh does not have a long history of a formal dairy marketing and processing industry compared with Kenya, but of course indigenous milk production, processing and consumption has ancient roots, particularly among the Hindu minority. From colonial times until the end of the 70's milk production and sale was predominantly done traditionally. Milk was mainly collected from small herds and sold in the nearby rural markets by the farmers.

More recently, simultaneous with the growing numbers of crossbred cattle in rural areas, middlemen and wholesalers have joined in collecting milk and selling to milk-sweet shops, curd, butter and ghee makers. The proportion of milk that goes to sweets and other products is estimated at around 42 percent (Survey). The major intermediaries in the milk trade are Gowalas, brokers (or Aratdar) and small processors.

Gowalas

The Gowalas collect milk from farmers, and transport it to other points in or near urban areas. Unlike mobile milk traders in Kenya, they also milk many of the cows from which the milk is collected. The price paid to the farmer thus includes a reduction to cover the milking service performed by the Gowala. These roles are often performed with help from family members. Hired labour involving three to five persons working for four hours each per day may also be engaged at times, to help with milking and transport. Some of the

Gowalas also have their own cows from which they collect milk for sale.

Brokers

Broker or "Aratdar" are legally recognised commission agents with a fixed establishment in an urban area where Gowalas assemble their milk for sale to other market intermediaries and processors. They receive a commission for every transaction, even though they may not physically handle the milk, rather just arrange the transaction. Though the brokers handle higher quantities of milk than considered by this study to be small scale, they are included because they are important facilitators in the small-scale milk trade. The broker may employ some three to five persons to help in unloading milk in the market. Their trade seems to involve little risk, as they do not incur costs in milk collection and transport. They provide market location and capital, in the form of premises, to support transactions between other agents.

Milk Processors

Milk processors in Bangladesh are involved in hot milk selling, milk sweet, curd and ghee (clarified butter) making. The milk sweet makers process and sell various types and flavour of milk sweets, which have a strong tradition in the local communities, especially for festivals. They purchase milk from farmers and Gowalas. Rural milk sweet makers mainly utilize family labour while the urban sweet makers frequently employ both family and wage labour.

The hot milk sellers prepare and retail the hot milk by the roadsides and frequently use family labour to do this. Some hot milk sellers do their business in rented or owned small shops. They procure milk from the Gowalas directly or through Aratdars. Curd making is mainly a rural based family business activity that is usually inherited. Men are involved in milk collection and sale of the curd while women do the actual processing. This sweetened form of curd is sold to traders who bulk and re-sell in urban areas, or directly to consumers, and cream is also sold to ghee makers. In circumstances of high demand such as festival seasons, hired labour is engaged.

The demand for curd decreases markedly during summer and monsoon seasons mainly because it can be substituted with fruits that are abundant in those seasons. Its demand is highest during winter. Ghee processors obtain milk cream from curd makers with whom they usually have a contractual arrangement. Some 10-12 curd makers would supply cream everyday to a ghee maker who processes about 55 kg from 110 kg of cream. The ghee processor engages wage employment for cream making, butter processing, heating, canning and packaging. Both casual and long-term employees are hired.

Ghana

The dairy industry in many parts of Ghana is relatively small and undeveloped. Populations living in the southern and coastal part of the country have little tradition of milk or dairy product consumption, as Trypanosomiasis and other diseases have historically restricted cattle keeping in those parts of the country. In the dryer north, however, cattle keeping is more common, although limited to indigenous breeds. With migration of northerners to the southern towns, and also the introduction of Western-style dairy products, dairy product consumption has grown.

The most commonly consumed products are condensed and sweetened condensed milk, much of it made from imported ingredients. There is still demand for fresh milk and indigenous products, however, especially wagashi, a soft cheese made traditionally in the Sahel region, and associated with the Fulani people now living in urban areas throughout Ghana. Much of the local market for raw and indigenous products targets the Fulani community as well as others who have acquired a taste for these products in spite of not having a tradition of their consumption. An important feature to note about milk supply in Ghana is that there is usually a separation between cattle owning and cattle keeping.

While cattle owners can be from nearly any ethnic group, they arrange to have the cattle kept and milked by Fulani cattle herdsman, who have greater indigenous knowledge of such practices. While the owners buy and sell the animals, the herdsman or stockmen collect and sell the milk. Within the families of some herdsman, some of the milk is processed into wagashi, generally by the women. Intermediaries in the milk trade in Ghana include assemblers, processors, and retailers.

The assemblers collect raw milk and wagashi (soft cheese) and deliver to the wholesalers and retailers in urban areas. Milk assemblers are mainly men while wagashi assemblers are female. The retailers are either mobile or sedentary traders. They mainly sell wagashi in small quantities. Small processors produce and retail cheese, ghee, yoghurt, fermented milk and ice cream. The stockmen double up in both roles as producers and retailers of milk. Stockmen's wives are generally the wagashi processors.

Milk Assemblers

Assemblers collect either milk or wagashi from rural producers and sell to retailers in urban areas. Milk assemblers are mainly men aged between 25 and 45 years while wagashi assemblers are women whose ages range from 18 to 55 years. They either collect the commodities or occasionally, the producers deliver them. The mode of transport is on foot and by public vehicles (bus or taxi). The assemblers engage both family and wage labour for the business activities.

Small Processors

Ghanaian dairy processors are mainly involved in the production and sale of soft cheese or wagashi in a cottage-industry type of production. Hard cheese, yoghurt and ice cream are also processed on a larger scale and in a more formal manner. Most of the wagashi processors are women, who are stockmen's wives. They collect and transport milk on foot, by bicycle or public transport, and also buy milk from their husbands. They engage family and wage labour in milk collection activities. Their business volume is influenced by seasonality of milk supply.

Retailers

Ghana's small-scale dairy retailers are individual dairy product sellers that deal with the sale of milk and processed dairy products. They are either mobile or sedentary. They buy their supplies from herdsmen, processors or assemblers. Though the smallest among the major agents in small-scale dairy marketing and processing, they are interspersed in many towns and offer employment mainly to family members.

Employment Opportunities

The role of employment in poverty-reduction programmes in developing countries has received considerable attention worldwide, in development strategies and policies. Many new employment opportunities in many developing countries are created in the informal sector whose rate of growth may be higher than that of the formal sector. For example, in Kenya, according to a recent economic survey the informal sector has been growing at over 10 percent in the last decade and its share of total employment, excluding employment in small-scale farming activities, was estimated at 70 percent in 2000. In contrast, the number of wage employees in the formal sector has remained static.

Despite continued rural to urban migration, a large proportion of the population in many developing countries still lives in rural areas and are mainly poor with over 30 percent overall classified as poor. Kenya, Bangladesh and Ghana are typical cases. The rural population in the three countries accounts for over two-thirds of the population and the rural labour force is growing at about 3 percent annually in each country. This increasing labour force in developing countries may not be absorbed productively in on-farm work given the limits to arable land, and increases in agricultural labour productivity through technology that reduces demand for labour.

Employment opportunities in rural areas may have to rely on strengthening the ability of non-farm agricultural activities to absorb the labour. Diversification into non-farm activities constitutes on average about 45 percent of rural incomes in developing countries and the "push and pull" factors driving this diversification are bound to persist. Push factors include changes in technology in agriculture that require less labour, creating labour surpluses and reducing agricultural labour opportunities, and pull factors include job creating in urban areas from industry that raise wages and employment opportunities there.

Dairy markets offer good opportunities for non-farm rural and urban employment. The perishable nature of milk, as well as the wide range of products that it can be transformed into, presents opportunities for value addition through the use of labour. Its relatively high unit value makes the use of labour for handling and processing economically viable. This may particularly be true in informal milk markets, which rely less on modern milk processing equipment, and more on traditional labour-intensive technologies. The case studies described here shed some light on the nature and quantity of employment created through small-scale dairy marketing and processing, most of which occur in the informal sector.

Kenya, Bangladesh and Ghana are chosen for the case studies because they represent a range of market-oriented dairy production systems, market channels, and of traditional dairy product consumption habits from the developing world. The cases thus provide an opportunity to gain strategic insights into how small-scale dairy marketing and processing can contribute significantly to rural and urban employment, and into the policies and training activities needed to support that process.

The proportion of people reported unemployed ranges from 20 percent in Ghana to 50 percent in Kenya. Furthermore, most of those people are apparently located in rural areas. The analyses are mainly based on data collected in April and May 2001, as well as on data gathered in previous milk market studies by ILRI and national partners in 1999 and 2000, through the Smallholder Dairy Project in Kenya (SDP), and the

Milk Marketing, Processing and Public Health Project in Ghana.

In the next sections, a conceptual framework depicting the linkage between employment, poverty and food security is presented, and then a framework for analysis of employment effects. This is followed by a description of dairy marketing and processing in Kenya, Ghana and Bangladesh within their respective historical contexts; the types of small-scale dairy marketing and manufacturing enterprises; survey results; and, policy recommendations for the future based on the findings.

Employment Generation

The use and value of employment in targeting poverty is neither new nor is it restricted to any one region. Historically, the Poor Employment Act of 1817 in Great Britain represented a major milestone in the development of economic policy to reduce poverty through employment and development. That Act and similar public interventions in later years in various countries have been largely based on the recognition that small firms are important generators of employment.

Hughes in interpreting the job-generating role of the small enterprises emphasizes the extreme skewness and volatility of the individual small business growth patterns, and the low quality and sustainability of the many jobs created by the mass microenterprises. He observed that a relatively few firms exhibiting rapid and sustained growers account for the bulk of sustained job generation in small firms. This is especially so for the developing world where economic reforms are recent and hurried, and availability and access to market information is not uniform and assured. Those small enterprises that are better placed to access resources including information and technology thrive and offer sustained jobs while their disadvantaged contemporaries fizzle out.

The development and growth of small enterprises is pegged to the existence of some level of entrepreneurial climate or enterprise culture amongst the people. While entrepreneurial skills exist in all cultures, they may vary in degree according to traditions, environment, and history. Such skills and motivations should be supported by a well-defined institutional structure that is understood by the participants, and which includes formal rights and protections to physical and other property. Access to resources in the form of capital, labour and infrastructure will then lead to the development of small enterprises participating in marketing and/or processing of farm produce.

As Reardon *et al.* point out, household members will redirect their labour away from land-based activities with the existence of:

(i) pull factors such as higher incomes in the non-farm sector relative to the farm sector; and

(ii) push factors such as increase in agriculturally sourced risk (farming that cannot ensure year-round income and consumption).

The main goal behind enterprises is to generate income for the entrepreneurs and their families. In the case of marketing enterprises, that requires that goods have to be sourced, transported, transformed, and marketed. The physical assets involved need to be acquired, serviced and repaired. All these activities require human labour fully or in part, depending on the level of business sophistication and technology employed. Human labour has a price, whether it is family labour (referred to as self-employment) or non-family labour (wage employment).

The remuneration received by family members is critical to a household's ability to access basic needs, improve livelihoods and create assets. As explained in a recent World Bank publication, the role of private enterprise in poverty reduction has long been neglected by development planners, researchers, and governments. This is supported in the article by several studies, showing first that economic growth has consistently led to reduction in poverty, and secondly that the engine to economic growth is private enterprise. The key to poverty reduction is thus demand for employment driven by economic growth.

Employment is categorised here in two broad categories: agricultural and non-agricultural. Agricultural employment is mainly rural/land based crop and livestock farming, including fish culture, and rural employment generated thereof in marketing and processing. Non-agricultural employment on the other hand is defined as the mainly urban based largely encompassing service subsector employment, formal and informal manufacturing and large-scale enterprises. There is often a close linkage between the two broad categories especially in developing countries where non-agricultural activities play a major role in complementing the activities of agricultural marketing and processing enterprises.

Employment generated by agricultural marketing and processing incorporates those labour-using activities that procure, process and market farm produce. With regard to dairy marketing and processing, the agents involved comprise raw milk traders, dairy processors and traders in processed dairy products. It should be noted, however, that these agents also hire services from others to repair their equipment, provide transport, etc., which in turn provides indirect employment from the dairy enterprise.

As indicated in the countries being addressed, much of the employment created in processing and in marketing occurs in the informal sector, whether rural or urban. Data on the informal sector compiled by the International Labour Organisation (ILO) indicates that for the poorer economies of the world, the informal sector provides more than 50 percent of total employment. Among the countries with the highest proportion of informal sector employment are two included in this study, Kenya and Ghana, with more than 50 percent and 70 percent, respectively, of informal employment as share of total employment.

Further, in nearly all those countries where data were available, 17 countries including Kenya, the share of employment in the informal sector grew during the 1990's. The key point here is that any attempt to increase employment must include emphasis on the informal sector, largely made up of individual entrepreneurs or small-scale enterprises. In the case of milk and dairy products, the existence of large informal markets for raw milk or traditional products is based on several key factors: the unwillingness of resource-poor consumers to pay the additional costs of pasteurisation and packaging, and the preference for traditional products, including raw milk.

REFERENCES

Gerber, M.E., *The E-Myth Revisited: Why Most Small Businesses Don't Work and What to Do About It*, HarperCollins, New York, New York, 1995.

13

Integrated Dairy Farm Management

Integrated dairy farm management means nothing more than beginning to view your dairy farm as more than just the sum of its parts.

PLANNING

A common producer response to adversity is to pull the hat down a little tighter, get out there and work a little harder and a little longer. However, many producers are working as hard and as long as they can. They have already been cinching down the belt, trimming here and there to achieve that elusive "efficiency." Planning is the tool that allows the technician to both free the entrepreneur to be visionary and to harness the manager through development of systems. A good strategic plan should allow you to work smarter instead of just harder.

Planning forces you to anticipate problems and take steps to resolve them. You must have an inventory of your equipment, its age and use. You must have projected farm income and expense for the year. Equipment purchases are somewhat long-term decisions. But contrast knowing your projected annual equipment purchases to the common way in which these decisions are made.

As a technician, you are faced with a seemingly endless "to-do" list and you must concentrate on the present. But as a proactive manager, it is your responsibility to look around and determine if you're labouring effectively. Are you using the best, most affordable technology? Are the most important tasks getting done? For example, are the maternity pens cleaned regularly, or has this job taken a back seat because you are too busy treating toxic mastitis cases? The second job is easily more urgent than bedding or cleaning pens. This simple example illustrates just how easy it can be to keep very busy while at the same time accomplishing very little.

Planning establishes a clear direction for both management and employees to follow. Many producers groan when they hear the words "mission statement." One slightly cleaned-up description is "a piece of paper on the wall that gathers dust." Indeed, plans need to be constantly reviewed and changed. General Eisenhower said, "The plan is nothing, planning is everything." If planning occurs regularly, the mission statement can serve as a powerful filter against which decisions, both large and small, can be gauged to determine the appropriate answer for your particular farm business.

Planning defines in measurable terms what is most important for the business and helps allocate resources like land, labour, machinery, and equipment efficiently. Part of planning includes setting goals for things like financial efficiency, vacation time, production parameters, or home improvement. By setting these goals, managers have both created measures to allow monitoring of success and determined what issues are high priority and deserve additional resources.

Types of Planning

There are two main types of planning: strategic and tactical. Strategic planning is the big picture look. It describes the business's overall objectives and includes a mission statement, long term and short-term goals, and a tactical plan. The tactical plan, in contrast, is much more specific. It describes how your business will implement day-to-day operations to achieve its long-term goals and objectives.

Tactical planning is something producers do every day, though they may not realise it. When you decide on the day's priority tasks during morning milking or the coffee break, you have created a tactical plan. A tactical plan is important, but focusing solely on this portion of planning can make it very easy to end up in the "wrong jungle."

Planning Steps

Personal Mission Statement or Primary Aim

Planning begins by outlining who you are and what you want your life to look like. Gerber and others urge people to envision what they would like their funeral eulogy to sound like. This is a somewhat dramatic analogy. However, taking hold of your life and directing its outcome is a dramatic step.

Business Vision and Mission

A business vision statement lays out exactly what your business must look like for you to achieve your primary aim. The mission, or value, statement indicates what it is your business stands for. It is a touchstone through which all major decisions should be filtered.

Long- and Short-term Goals

Long-term goals are the first steps toward accomplishing your vision of your business. They should be achievable within 2-5 years and meet the following requirements. They should be: Directional and move you toward the objectives of your mission statement; Reasonable - practical and obtainable, not extreme; Inspiring, challenging, and affect you positively;. Visible - able to measure, easy to visualise; and Eventual - they will be fulfilled in the future.

Tactics

Tactics are the list and prioritisation of daily activities orchestrated to accomplish short-term goals. They are where most of us spend most of our time. A to-do list is a type of tactical plan. This is where all the work gets done - where the rubber meets the road. There is nothing wrong with the technician's work. The problems come when day-to-day work is not aligned to accomplish any higher purpose than to survive today. When tactics rule the business, the status quo is maintained, but goals are rarely reached.

Transitioning from a Tactical to a Strategic Mindset

Making a commitment to evaluate your mode of operations means stopping and thinking about what your primary aim is and then aligning your daily tasks with these goals. As you might guess, this can be very easily said, but much more difficult to do. Here are some ways to begin.

— *Start small.* Set aside one regularly scheduled hour with no disturbances to plan the upcoming week. If possible, plan the week in conjunction with relevant managers or family members. It is key that this hour is a high priority both to you and those involved. If you miss a few planning sessions or ignore the outcomes, your family and employees will follow your example.

  List what is going to come up in the next week. Be sure to include all facets of your life: farm, family, social, community, and others. Anticipate bottlenecks and crises. For example, could you run short of feed, labour, or supplies?

— *Prioritise the tasks.* This will help you stay focused on the important, not the urgent.

— *Set goals.* After a month or two of these weekly meetings, it may be easier for you to formulate goals. Some may be short-term and could be accomplished over the next few months. Others may be longer-term. If you have been able to involve key family members and employees, you may be surprised to learn about their priorities.

There are a lot of resources out there that understand you and your dairy farm. Use the ones that you trust to help you decide which tasks are important. Resources include your veterinarian, nutritionist, cooperative representative, Extension agent, and others.

Translate Strategic Plans to Tactical Plans

If you have accomplished the steps laid out above, you have begun to create an "umbrella plan" for your business. A plan that spans your entire operation and lays out a bold vision of your future. Accomplishing your plan will require the coordination of many different "departments" within a business. A traditional organisational chart generally can include departments of Research and Development, Manufacturing/Operations, Purchasing, Personnel/Human Resources, Public Relations, Sales/Marketing, Finance/Legal. Integrated dairy farm management implies that your strategic plan is translated through each department on your dairy into your tactical, or day-to-day actions. Each of these departments represents an important view of your farm.

Linking Financial Ratios to Production Decisions

Despite a growing recognition by dairy producers and their consultants of the importance of financial statements and their analysis, only some employ these tools on a regular basis. In an effort to make financial analysis accessible for both dairy producers and their consultants, a simple summary of production and financial measures has been designed. The output of financial analysis software can be lengthy. While this output is thorough and necessary as supporting documentation, it can lead to "analysis paralysis."

Finally, the summary includes previous years' analyses to facilitate trend evaluation. Figure 5 provides the layout of the summary along with standard calculations and, where available, guidelines for each measure. The calculation of each ratio follows guidelines recommended by the Farm Financial Standards Council.

Guidelines for and Analysis of Financial Ratios

Assessing financial ratios seems to reveal the competitive nature present in every human. "Is that good or bad?" and "How does that compare to other farm's numbers?" are the inevitable questions. For most ratios, the answer to the good or bad question is, "It depends." For those ratios that have anecdotally accepted ones, the guidelines are listed. When solid guidelines are not available, it is common to compare farm performance to that of other, similar farms. This is commonly referred to as benchmarking. Benchmarking can be a good source of information and motivation. However, it can be over used. If the records are available, it is often more useful to compare a farm to its own past performance, rather than to the performance of peers. Careful attention should be to the makeup of the peer group and method of calculation if benchmarking is to be undertaken. Finally, the analysis of the ratios presented below assumes a basic working knowledge of financial statements and their analysis.

Productivity

The summary begins by assessing farm productivity - setting the stage for the rest of the financial analysis. The critical dairy production parameters of milk sold per cow, total milk sold, milk price, and average cow inventory (herd size) are detailed.

Profitability and Net Worth

Through several measures, this section determines whether the farm is making money. The most traditional way this is measured is through Net Farm Income (NFI) - a return to unpaid labour, management, and equity. Net farm income per cow is simply NFI divided by average cow inventory. This number facilitates comparisons between farms of different herd sizes and is a number commonly found in lay publications. Acceptable ranges for NFI and NFI per cow are dependent on what part of the world the analysis is done in.

Return on assets (ROA) and return on equity (ROE) are alternative ways to measure farm level profitability. These measures subtract a charge for the value of unpaid operator labour and management from NFI and add back interest (in the case of ROA), then are divided by total farm assets or equity, respectively. Acceptable ranges for ROA and ROE depend primarily on individual investors. At a minimum, farming should return rates better than a savings account or

Certificate of Deposit at the local bank. Finally annual net worth on both a market and cost basis are included. Some consider annual net worth growth the ultimate measure of profitability. Farm level profitability is captured (or lost) through efficiencies gained in three main areas: asset use, operating, and labour use.

Asset Efficiency

Asset turnover (ATO) is the "purest" form of asset efficiency. While return on assets also may show how efficient the farm is at utilising their asset base, it is complicated by the necessary subtraction of an arbitrarily determined value of unpaid operator labour and management. Asset turnover is determined only by gross revenue, affected most by the price and volume of milk sold, and by farm assets. If this value is determined to be lower than optimal (40-60% for United States dairies), it is because volume or price of milk was lower than optimal, or because too many assets generated the volume of milk sold.

Examination of ATO reveals only farm-level area efficiency. If this can be improved, it may be useful to break down the farm by enterprises, or even by assets, to determine which are farm profit centres. However, this measure helps link the level of profitability directly to management practices. Does the producer value having large, new machinery? Is a large parlour used for a small- or medium-sized herd? Does the farm have enterprises besides dairy that require a large asset base or that have recently faced low yields or prices? If there are big differences in the level of cost and market ATO, which farm assets cause the discrepancy?

Operating Efficiency

The net farm income ratio (NFI%) determines what percent of gross revenue the farm is retaining as profit. Again, this number represents a farm-level assessment of efficiency. If this number is low (less than 15% on United States dairies), further examination is required. The operating expense ratio, depreciation expense ratio, and interest expense ratio can then be examined to determine what area(s) are contributing to a low NFI%. A high operating expense ratio indicates that one or more cash expenses are too high.

Purchased feed and hired labour generally represents the largest cash cost for dairy farms and are the first areas to examine. If purchased feed is high, how does it compare to previous years? Is the increase a trend or a blip? Did concentrate prices rise? Did a poor forage year result in the purchase of hay or silage? Were expensive additives included? "High" purchased feed costs can be acceptable if they result in high milk production. How does hired labour cost compare to previous years? Is the increase a trend or a blip?

Was a new practice, like timed breeding or 3x milking, instituted that resulted in more hired labour costs? Did the cost of hired labour in your area increase? If the depreciation expense ratio is high, it indicates that the farm may have too many assets for the level of profit. Are asset levels (cost basis) higher than on peer farms? Again, this could reflect the milk production or price conditions of a particular year, or it could represent an ongoing problem.

Labour Efficiency

Liters of milk sold per full-time employee (FTE) provides a method for examining labour efficiency. This calculation does not control for differences in labour needs across different farm types. Total labour hours, both paid and unpaid are also included to evaluate change across time. If labour efficiency is determined to be low, often parlour efficiency should be the next area examined.

Sometimes, farms will be quite asset efficient and at the same time have low labour efficiency. This is often due to a classic labour-for-capital trade-off. The machinery and parlour (or stanchion barn) may be old or lacking in the newest technology. However, output is maintained by additional labour hours, often contributed by the operator or family. If the dairy producer is near the beginning or ending of their career, this trade-off may be a necessary transition to time of easier cash flow or exit from the business. However, if this trade-off is taking place during the middle of the dairy producer's career, it is incumbent upon the analyst to make at least cursory inquiries about quality of life. Financial analysis is only a portion of whole-farm planning. Maximising profitability may not be the dairy producer's main goal.

Liquidity and Solvency

These values roughly measure the risk level that the farm can bear. The current ratio assesses liquidity, or the ability of the farm to meet short-term debt obligations. US bankers frequently use this calculation, but because of the seasonality of current assets (feed inventories, primarily) on dairies, it usually has little meaning. Debt-to-asset levels, on both a cost and market basis, are presented. A higher value implies higher risk. The lower this ratio, the more room a business has to utilise borrowed funds. A generally acceptable range for United States dairies is 40-65%. As the ratio increases, interest expense becomes larger. This can lower profitability (NFI).

In addition, the large payments incurred by the debt create a need for consistent and predictable cash flows. Higher debt levels means that the dairy may not be able to withstand a costly event such as decreased milk production from a disease outbreak or ration imbalance. Total debt is provided so the farm can track trends across years.

Towards Successful Business Management

We have seen many changes in the dairy and other agricultural industries in recent years. Milk prices are more volatile. Costs of the various inputs continue to rise. As herd sizes increase, more dairy producers are hiring employees to provide needed labour. To succeed in this business, dairy producers must be willing to consider new ways of doing business.

Successful management of a dairy farm business involves optimum use of available resources -land, labour and capital. How to make best use of the business's resources will vary because each farm has its own unique set of resources. Dairy producers must be willing to consider some non-traditional or innovative alternatives in handling various aspects of their business. Many factors contribute to the success of dairy farms. There are several characteristics that many successful dairy farms share.

Goal Setting

Establishing goals is an important first step for a dairy farm business. Goals are used to guide the decisions made for the business and to measure their results. Goals are considered when deciding the most appropriate use of the farm's resources.

Many of the farm's goals will focus on profitability of the business. However, other goals may be just as important as earning money. Dairy producers want to be well respected by their family, friends and neighbours. They also want to make a positive contribution to their communities, including using environmentally sound farming practices and being good stewards of the land. Who should be involved in the goal setting process? All family members actively involved in the dairy operation need to work together in determining the long-term goals of the business.

Short-term goals may be developed by those individuals who have direct responsibilities for a specific aspect of the operation (e.g., crop production, milking herd, or heifers) with a review by all family members before the goals are finalised. Involving key employees in establishing short-term goals and communicating the business goals to all farm employees will help them to understand the business and be more committed to their jobs.

Decision Making Process

Dairy producers make many decisions throughout the year about their business. Some require minimal effort while others represent significant consequences to the farm. For these major decisions, considerable time is spent gathering information and analysing the options. Following a systematic approach to decision-making will ensure that the manager makes logical, information-based decisions. The steps in the decision-making process are:

1. Specify an objective or goal
2. Identify and define the problem(s)
3. Collect data and information
4. Analyse various options for taking action
5. Make the decision-select the best option
6. Implement the decision
7. Monitor and evaluate the results of the decision
8. Accept the consequences of the decision

Effective Utilisation of Information

This is the information age. A multitude of information is available from many sources. Avoiding information overload can be a challenge. During the decision making process dairy producers also may wonder if they have the most appropriate information.

Some dairy producers are working with management or advisory teams to improve farm productivity and meet their goals. Producers bring together various people who already work with the farm, such as the veterinarian, feed consultant, lender, milk cooperative representative, crop consultant, and extension agent. By working as a team these individuals offer different perspectives in evaluating a problem and determining the alternative solutions. Many times more viable alternatives are identified than would have been considered by any one of the team members alone.

Cost of Production

Most dairy farm businesses have several enterprises within the operation, such as the milking herd, replacement heifers, forage production, and other crop production used as feed. By keeping detailed financial records, producers can determine the costs associated with each enterprise and decide the most profitable alternative. In recent years some dairy producers have decided to contract services associated with all or part of a specific enterprise.

Human Resource Management

Employees are the most important asset on dairy farms. Employee management may represent the greatest challenge on dairy farms. Different skills are needed to effectively work with employees compared to overseeing the milking herd or cropping programme. Dairy producers must possess or develop excellent communication skills to be successful employee managers.

Frequent communication with employees is critical. Employees need to clearly understand the expectations of their jobs. They want to know how they are doing in their jobs. They also need to express any concerns or suggestions about their work.

Many effective forms of communications exist on dairy farms. On farms with one or two employees, one-on-one informal visits may be most beneficial. Some farms are holding regular staff meetings to update employees and obtain feedback.

References

Kay, Ronald D. and William M. Edwards, *Farm Management*, McGraw-Hill, Inc. New York. 1994.

White, Gerald, "Using strategic planning to formulate future business opportunities", *Cornell Agricultural Economics Staff Paper*, Cornell University, Ithaca, New York, 1989.

Bibliography

Anon, *The grass farmer's bookshelf,* The Stockman Grass Farmer, June 2001.

Arora, Dinesh ed, *Biotech's Dictionary of Dairy Science,* Biotech Books, New Delhi, 2006.

Atkins, P.J., "The pasteurization of England: the science, culture and health implications of milk processing, 1900-1950", In: Smith, D. & Phillips, *Journal of Food, science and regulation in the 20th century,* Routledge, 2000.

Barnhart, Stephen K., *Selecting Forage Species,* University Extension, Iowa State University, Ames, 1999.

Berg, K., *Houses for self recruiting beef cattle, Norway,* 1996.

Cielejewski, H., *Experience with cold housing for dairy cows, Landtechnik,* 1997.

Cobleigh, R., *Handy farm devices and how to make them,* New York: The Lyons Press, 1996.

Cooper, M.M., *Beef Production,* Biotech Books, New Delhi, 2005.

Davidson, J., & Chase, L. *Farm machinery: Practical hints for handy-men,* New York: The Lyons Press, 1999.

Eckles, Clarence H & Ernest L Anthony, *Dairy Cattle and Milk Production,* Biotech Books, New Delhi, 2007.

Engle, Cindy, *Wild Health: How Animals Keep Themselves Well and What We Can Learn From Them,* Houghton Mifflin Company, New York, NY., 2002.

Fiez E., "Sizing up heifer age", *Dairy,* January, 1989.

Fisher, A.D, Crowe, M.A, Prendiville, D.J, Enright, W.J., "Indoor space allowance: effects on growth, behaviour, adrenal and immune responses of finishing beef heifers", *Animal Science: An International Journal of Fundamental and Applied Research,* 1997.

Gegner, L., "Value-Added Dairy Options", *ATTRA* (Appropriate Technology Transfer for Rural Areas), Fayetteville, Ark., Aug. 2001.

Gerrish, Jim, *Strategies for pasture improvement*, Forage Systems Update, January 1999.

Ghosh, Bibek ed, *Crops Residues and By-Products in Animal Feeding*/ Genetech 2007 Rs. 795 Rs. 795

Godbole, N.N., *Milk: The Most Perfect Food*, Biotech Books, New Delhi, 2007.

Gunby, L., *Early farm life*, New York: Crabtree Publishing Company, 1992.

Gunnink, Doug, *Gross margin analysis helps show the way to grazing profits*, The Stockman Grass Farmer. April 1993.

Harvey, W.M. Clunie & Harry Hill, *Milk Products 2nd edn*, Biotech Books, New Delhi, 2006.

Heinrichs A.J, et al, "Standards of weight and height for Holstein heifers", *Journal of Dairy Science*, 1987.

James R.E, et al, "Heifer feeding and management systems", In H. Van Horn and C. Wilcox, (eds): *Large Dairy Herd Management*, American Dairy Science Association, Champaign, Ill. 1992.

Jull, Morley Allan, *Successful Poultry Management 2nd edn*, Biotech Books, New Delhi, 2007.

Kutty, C. Ibraheem & Sheeba Khamer, *Milk Production and Processing*, Daya Publishing House, New Delhi, 2004.

Leep, Rich, and Doo-Hong Min., *Clovers beat commercial N in Michigan studies*, The Forage Leader, Spring., 2005.

Levantin, D.L., "Developing beef cattle husbandry", *Zootekhniya*, 1996.

Mader, T.L, Hahn, G.L, Gaughan, J.B., Dahlquist, J.M. "Shade and wind barrier effects on summertime feedlot cattle performance", *Journal of Animal Science*, 1999.

Marwaha, Kavita ed, *Meat Hygiene*, Genetech Books, New Delhi, 2007.

Milk Industry Foundation, *Analysis of Milk and Its Products: A Lab Manual 2nd edn*, Biotech Books, New Delhi, 2005.

Murphy, Bill, *Greener Pastures On Your Side of the Fence*, Arriba Publishing, Colchester, VT., 1987.

Nasirov, U., "Animal husbandry system, its status and development problems in Uzbekistan", *Dryland Pasture, Forage and Range Network News*, 1996.

Okularczyk, S, Novak, T, Sowula, E., "Cows and beef cattle handling costs with different production scale and mechanization levels", 1996.

Owen, W. Powell, *Poultry Farming and Keeping*, Biotech Books, New Delhi, 2007.

Pang, H, Makarechian, M.H, Basarab, J.A., "A simulation study on the effects of cow size and milk production on bioeconomic efficiency of range beef cattle", *Journal of Applied Animal Research*, 1999.

Paul, Meenakshi ed, *Formation of Bioactive Peptides from Dairy Products*, Genetech Books, New Delhi, 2007.

Pegler, Henry Stephen Holmes, *Goats and their Profitable Management 9th edn*, Biotech Books, New Delhi, 2005.

Rao, V M, *Evaluation of Women Dairy Project in Rajasthan*, Regency Publications, New Delhi, 2003.

Rath, S., *About cows*, Minocqua, WI: Heartland Press, 1987.

Richmond, H. Droop, *Laboratory Manual of Dairy Analysis 3rd Revised edn*, Biotech Books, New Delhi, 2004.

Ruoho, O., "Elimination of Salmonella infection in a beef herd in a loose housing system", *Suomen Elainlaakarilehti*, 1996.

Singh, Vir & Babita Bohra, *Dairy Farming in Mountain Areas*, Daya Publishing House, New Delhi, 2006.

Sonnenberg, H., "Straw comfort: mechanical straw preparation for loose litter livestock husbandry", *Landtechnik*, 1998.

Suzuki, K, Oizumi, C, Kobayasi, M, Mori, T, Wakamatu, M, Sajiki, S., "Effect of the wind blowing right below down by the fan for the beef cattle-house", *Bulletin of the Chiba Prefectural Livestock Experiment Station*, 1999.

Turner, Newman, *Fertility Pastures and Cover Crops*, 2nd ed. Bargyla and Gylver Rateaver, Pauma Valley, CA., 1974.

Wassmuth, R, Wallbaum, F, Langholz, H. J., "Outdoor wintering of suckler cows in low mountain ranges", *Livestock Production Science*, 1999.

Index